MW00957570

How to make quality biodiesel at home

A simple step-by-step biodiesel manual & guide

By

P Xavier BSc MSc

Copyright © 2020

ISBN 9798674439554

Acknowledgements

I'd like to thank Richard McInnes, Jim Scutt, Vanessa Thompson & everyone at Cedar Garage Worthing for their excellent & invaluable help with this book.

Introduction

Imagine the satisfaction you will feel knowing that the fuel you've made at home allows you to drive your vehicle, or heat your house at a reduced cost whilst also reducing the environmental impact. Biodiesel production allows you to save both money & save the planet too. This book allows you to achieve all this with simple step by step instructions. It starts simply giving a background to diesel, the diesel engine & therefore why vegetable oil needs to be altered slightly to allow it to power a modern diesel engine. Then simple recipes allow you to produce larger & larger batches that will you can use to run your vehicle & heat your house using the biodiesel that you have created.

All relevant areas are explored, such as, is it possible to use straight vegetable oil in a diesel engine, is it possible to use waste engine oil to power a diesel engine & how to produce the best quality biodiesel for use in your own vehicle.

Instructions are given in a simple step by step format. Starting with a trial mix using clean vegetable oil, then stepping up to using waste vegetable oil & ultimately up to a large scale home production method. It is always advisable to start small & master that method before stepping up to the next phase. All methods, materials, equipment & safety considerations are listed & explained throughout, enabling the reader not only to produce quality biodiesel, but to do so safely & knowledgeably time after time.

This book is a biodiesel manual that you will refer to again & again whilst you create your home made biodiesel. All temperatures are given in imperial & metric, allowing the manual to be easy to follow regardless of where you live or whatever system of measurement you are most comfortable with.

Table of Contents

ACKNOWLEDGEMENTS 2

INTRODUCTION 3

TABLE OF CONTENTS 4

NOTES 10

CHAPTER 1 – THE MOTOR VEHICLE 11

A BRIEF HISTORY OF THE MOTOR VEHICLE 11
THE DIESEL ENGINE 15
DIESEL DESIGN 19
DIESEL INJECTION 23
TURBO CHARGING & SUPER CHARGING.......................................26
DIESEL PARTICULATE FILTER & CATALYTIC CONVERTERS...............28
LAMBDA SENSOR 33

CHAPTER 2 – DIESEL FUEL 34

CRUDE OIL..34
HYDROCARBONS...35
FRACTIONAL DISTILLATION 36
FLAMMABILITY & VISCOSITY 38
FORECOURT DIESEL 40

CHAPTER 3 – ORGANIC CHEMISTRY MADE SIMPLE 43

CHEMICAL LEGO 43
CHEMICAL CHAINS..44
CHEMICAL FAMILIES..47
CHEMICAL GROUPS & CHEMICAL BONDS 48

CHAPTER 4 – SOURCES USED FOR FATTY ACID METHYL ESTERS
(FAME)..52

PLANT DERIVED FAME'S 53
  Arachis hypogaea – peanut...53
  Brassica napus – rapeseed...54
  Cannabis sativa – hemp...55
  Carthamus tinctorius – safflower 56
  Cocos nucifera – coconut...57
  Elaeis guineensis – oil palm 58
  Glycine max – soybean 59
  Helianthus annus – sunflower 60
  Jatropha curcas – jatropha...61

Zea mays – corn 62
ANIMAL DERIVED FAME'S 63
ALGAE DERIVED FAME'S 65
CONCLUSION 67

CHAPTER 5 – ENVIRONMENTAL CONSIDERATIONS 68

THE COMPOSITION OF PETROCHEMICAL DIESEL POLLUTION 68
THE COMPOSITION OF STRAIGHT VEGETABLE OIL (SVO) POLLUTION 72
THE COMPOSITION OF BIODIESEL POLLUTION 75
REDUCING POLLUTION FOR PETROCHEMICAL DIESEL, STRAIGHT VEGETABLE OIL (SVO) & BIODIESEL 76
EVEN MORE SAVINGS 82

CHAPTER 6 – MODIFICATION OR TRANSFORMATION 85

ENGINE MODIFICATION 85
ELSBETT ENGINE 90
FUEL MODIFICATION 91
Base-catalysed transesterification 91

CHAPTER 7 – DANGER, DO NOT TOUCH 94

METHANOL 95
First aid for methanol 97
SODIUM HYDROXIDE 98
First aid for sodium hydroxide 100
POTASSIUM HYDROXIDE 101
First aid for potassium hydroxide 102
SULPHURIC ACID 103
First aid for sulphuric acid 104
GLYCEROL 105
First aid for glycerol 106
ISOPROPYL ALCOHOL 106
First aid for isopropyl alcohol 109
PHENOLPHTHALEIN 110
First aid for phenolphthalein 111
BROMOPHENOL BLUE 111
First aid for bromophenol blue 112
HYDROCHLORIC ACID 113
First aid for hydrochloric acid 115
ACETIC ACID 116
First aid for acetic acid 118
GENERAL SAFETY 118
General chemical advice 119
Oil advice 120
Heat advice 121
Electrical advice 121
PPE advice 121
Other sources 122
Other chemicals 123

CHAPTER 8 – EQUIPMENT 125

SAFETY EQUIPMENT 125
Eye wash 125
First aid kit 126
Fire extinguisher 126
Fire blanket 128
Shower hose 128
Spill kit 129

PERSONAL PROTECTIVE EQUIPMENT (PPE) 129
Gloves & gauntlets 129
Coveralls 132
Footwear 134
Goggles or face protection 136

MISC SAFETY EQUIPMENT 138
Mobile phone 138
Thermal imaging camera 139
A N Other 140
Vinegar 140

LABORATORY & USEFUL EQUIPMENT 141
Hand transfer pump 141
Weighing scales 141
Blender 142
Pasteur pipette aka eye dropper (graduated) 142
Hydrometer 143
Thermometer 143
Borosilicate laboratory beakers 143
Borosilicate Petri dishes 144
Borosilicate stirring rods 144
Electronic pH meter, litmus paper & phenolphthalein 144
Laboratory test sieves 145
Electric hotplate 145
Plastic home brew buckets 146
Stainless steel cooking pot 146
Pen & notepad 146
Small lab spoons 147
Preserving jars 147
Brushless drill 147
Mixing paddle 148
300ml tin 148
Stopwatch or watch with a second hand 148
Ladle 149
In-line fuel filters 149
2m fuel hose 149
Small funnel 149
Biodiesel storage container 150
Flexi tubs 150
Petrochemical diesel (sample) 150

LARGE SCALE PRODUCTION EQUIPMENT 150
    Fume cabinet 151

CHAPTER 9 – THE FIRST BATCH 152

THE TEST SOLUTION 153
TEST BATCH №1 – SVO 153
    The equipment 154
    Step 1 – Safety first & de-watering 155
    Step 2 – Titration, titration, titration 156
    Step 3 – Methoxide magic 159
    Step 4 – The big mix of compromises 160
    Step 5 – Separation 161
    Step 6 – Testing 163
    Step 7 – Washing the biodiesel 164
    Step 8 – Filtering & de-watering 166
    Step 9 – The clean up operation 168
TEST BATCH №2 – WVO 169
    The equipment 170
    Step 1 – Safety first & de-watering 170
    Step 2 – Titration, titration, titration 170
    Step 3 – Methoxide magic 171
    Step 4 – The big mix of compromises 171
    Step 5 – Separation 171
    Step 6 – Testing 171
    Step 7 – Washing the biodiesel 171
    Step 8 – Filtering & de-watering 171
    Step 9 – The clean up operation 171
CONCLUSION 172

CHAPTER 10 – SCALING UP PRODUCTION 173

LONGEVITY OF THE BIODIESEL 173
A TAXING CONUNDRUM 173
    Planning permission 176
    Environment Agency 177
    The Health & Safety Executive (HSE) 177
    Insurance 178
    Trading Standards 178
A BIG WASH 179
    Static water washing 179
    Mist washing 180
    Pump washing 180
    Bubble washing 182
    Drying biodiesel 183
SCALE OF SETUP 185
    Self build, own design 185
    Self build, others design 188
    Off the shelf, turn key system 190
    Checklist 190

Storage 198
Design thoughts... 200

CHAPTER 11 – CLEANING UP 202

WASTE WATER.. 202
  Methanol recovery 202
  Soap & grease recovery 203
  Dealing with the potassium hydroxide............................ 204
WASTE GLYCEROL... 205
  Methanol recovery 205
  Dealing with the potassium hydroxide............................ 206
  Soap & grease recovery 206
COMPOSTING THE GLYCEROL... 207
BURNING THE GLYCEROL 208
DUST SUPPRESSION WITH GLYCEROL 209
SOAP MAKING WITH GLYCEROL.. 210

CHAPTER 12 – WINTER BIODIESEL 212

WINTER MOTORING 212
  Measuring the cloud point & gel point 213
  Reducing the cloud point... 215
  Block heater... 215
  Battery heater.. 216
  Engine heater 217
  Fuel tank heater... 217
  Fuel filter heater.. 217
  Fuel additives 217
  Further measures 218

CHAPTER 13 – WHAT WENT WRONG?.......................... 219

MY BIODIESEL DOESN'T LOOK RIGHT. WHAT'S GONE WRONG? 219
  Emulsion looking like mayonnaise 219
  Thick gel looking like pudding or jelly........................... 221
  Oil & glycerol not separating out 222
  There's a soapy layer in the batch 223
  Solids in the biodiesel... 224
  Acidic biodiesel... 224
  How do I know if the biodiesel is clean enough to use in my car?..225
THE FUEL FILTER IN THE ENGINE KEEPS GETTING BLOCKED. WHAT'S WRONG
WITH THE BIODIESEL TO CAUSE THIS? 225
I'M LOOSING ENGINE POWER WHEN I USE BIODIESEL. IS THIS NORMAL? 226
THE WARNING LIGHTS HAVE COME ON IN MY VEHICLE WHEN I USED THE
BIODIESEL. IS THIS NORMAL? 226
CAN I ADD PETROL TO MY BIODIESEL TO HELP IT RUN IN WINTER?.............227
CAN I JUST USE OLD ENGINE OIL TO FUEL MY DIESEL ENGINE? 227
HOW PURE IS THE GLYCEROL BYPRODUCT? 227

CHAPTER 14 – VARIOUS TESTS.................................... 229

The level of conversion test..229
The cloud point & gel point test 230
The HMPE filtration test..230
The FFA test...231
The methanol test - purity..232
The pHLip test 233
The separation test...233
The take away titration test..234
Tests for water in WVO 234
Titration of WVO 237

CHAPTER 15 – ALTERNATIVE BIODIESEL RECIPES 238

SINGLE STAGE RECIPES 238
The standard biodiesel recipe 238
The high yield biodiesel recipe 239
The high titration adjustment factor recipe 239
TWO STAGE BASE RECIPES 240
The 80 – 20 recipe..240
The base base biodiesel recipe...240
The zero titration two stage recipe 241
TWO STAGE ACID RECIPES 242
The FATTA recipe 243

CHAPTER 16 – GLOSSARY OF TERMS 244

CHAPTER 17 – EUROPEAN BIODIESEL 259

TABLE OF ILLUSTRATIONS..260

INDEX..263

ABOUT THE AUTHOR 279

Notes

This work is by the author & as such, all copyright belongs to the author. You are not permitted to copy any text or images without the authors express permission.

Some of the images in this book are created by the book's author, others are either in the public domain or individually credited to their respective creator &/or copyright owner. Any other images are from sources where they are copyright free.

This work is fully referenced to aid the reader in any future studies. That being said, internet references have been provided to aid study as most individuals do not have huge libraries at their disposal. It is far easier to research material online than to go through the expense of ordering specific books at a public lending library.

A word of warning. As making biodiesel involves the use of volatile & dangerous chemicals, it is advised that stringent safety procedures should be followed at all times. It is also advised that ALL relevant safety equipment & PPE are used when handling or using these dangerous chemicals. All steps & storage should be undertaken outdoors to minimise the risk of fire & accidental inhalation of fumes. Also, if defective biodiesel is manufactured or not cleaned sufficiently, it can cause damage to a diesel engine. Due to these facts, the information contained within this book should therefore be treated solely as information & as such, the author accepts no liability for any accidents or damage that may occur whilst making biodiesel, or for any resultant damage from using defective biodiesel.

Chapter 1 – The motor vehicle

The reasons you are reading this book is because either you have an interest in saving money, saving the planet or wish to future proof your transport arrangements for the coming years as it is highly likely that traditional petrol/diesel fuelled vehicles will become outlawed at some point.

Therefore, it would be advantageous to understand exactly how you can continue to drive your diesel vehicle safely, legally & save the planet too. Even with just a basic level of comprehension on biodiesel will allow the reader to make more informed choices in the future.

Also, if you think your diesel vehicle can not run on biodiesel, it may come as surprise for you to know it already does in part, as the fuel you currently purchase at the forecourt already contains a percentage of biodiesel. It therefore makes sense to understand why.

The best place to start is to understand where diesel comes from & how a diesel engine works. It will then be easier to understand how petrochemical diesel fuel can be replicated & why.

A brief history of the motor vehicle

The internal combustion engine is acknowledged to have been invented in 1861 in Germany by Nikolaus August Otto[1]. At that time Otto was enhancing an earlier engine built in France by Jean Joseph Etienne Lenoir which used gas as a fuel.

---

[1] https://en.wikipedia.org/wiki/Nikolaus_Otto - 26/02/2019

Otto worked to develop it as a liquid fuel (petrol) engine & what he created became known as the Otto engine & the principles he developed in his engine (known as 'four stroke') are still used to this day in most engines. In the following years, many individuals worked on a variety of engines that used a variety of fuel sources. Electric, hydrogen, petrol & steam were just a few. Then in 1892, Rudolph Diesel was granted a patent for his New Rational Combustion Engine. By 1897 he had created the first diesel engine.

It should be understood that the development of engines at this early period was widely dispersed, took place over a wide timescale & utilised numerous power sources. This was all due to the fact that each of the inventors utilised the power sources that they had to hand. Steam, electric & petrol were therefore all competing head to head for dominance in this fast emerging engine market. But it was not until the network of petrol filling stations were built shortly after 1910, when the Ford Model T became popular, that the need for petrol stations were realised & that is how petrol fuel came to dominate the market as a vehicle fuel. This is the only reason why the petrol fuelled internal combustion engine achieved dominance over all its competitors as until these petrol filling stations were built, petrol was only obtainable from chemist shops.

Over the preceding years, enhancements were then developed for the petrol fed internal combustion engine, then enhancements in driveability & safety of the vehicle quickly followed. WWI was a particularly active time for innovation, but it was only during the oil crisis of 1973 when the development focus started to move seriously into other technologies. Since 1892, the diesel engine was quietly being developed alongside petrol & now 20% of all passenger cars globally are diesel powered, with Europe now having 47% of its passenger cars being diesel powered. The uptake of diesel powered vehicles is currently still increasing globally, most prominently in India, Japan & South Korea.

However, the majority of the worlds large motor vehicles such as vans & lorries are diesel powered vehicles, along with nearly all mobile plant (such as diggers) & also many of the world's trains. The infrastructure of the developed world is therefore diesel powered. When materials are dug from the ground, diesel power is used. This material is then moved to processing & refinement sites using diesel power. Those refined materials continue to be moved & during every step, it is always diesel power that moves these products, from source to manufacturing, to shops & finally to the consumer. Public transport is also predominantly diesel powered. If the majority of the world's cars are fuelled by petrol, it would be safe to say that the entirety of the world's infrastructure is fuelled by diesel.

In addition, what most people do not realise is that the majority of the worlds electricity is also derived from burning fossil fuels, therefore even electric vehicles can be said to be in part, diesel powered. In 2016, 65.3% of all electricity in the world was obtained by burning fossil fuels. Every country has different resources, so each country has different sources for their electrical generation. France uses a high proportion of nuclear fission, therefore only 10% of its electricity is produced from burning fossil fuels. However in the USA, 70% of their electricity is generated from fossil fuels, whilst in China 80% of their electricity is made from burning fossil fuels.

You may be thinking 'why has the world become so reliant on petrol/diesel power?' The simple answer is that petrol/diesel (any oil derived fuel) has more extractable energy per kilo than any other commonly available fuel source. The simple graph in figure 1 demonstrates this simple fact while comparing it to various other power sources, only hydrogen has more extractible energy.

At the time of writing this book, the cost on the UK for diesel is £1.10p per litre & petrol being £1.00p per litre.

The equivalent amount of hydrogen currently costs £9.99p, therefore despite hydrogen containing three times more extractable mega joules per kilo making it more energy dense, it is currently 10 times more expensive than petrol, which does not allow it to be an economically viable alternative to petrochemical fuels.

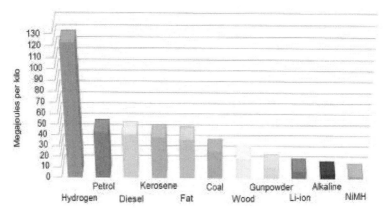

Figure 1 Extractable energy from various fuels per kilo (P Xavier ©
2019)

Despite petrol/diesel being the only viable power source for vehicles, both the UK & the EEC governments have announced that they now intend to legislate & outlaw the sale of all new diesel & petrol fuelled vehicles by 2040[2].

This has predictably caused problems for all vehicle manufacturers, mechanics & anyone who may wish to purchase a new vehicle, as a credible replacement to the petrol or diesel powered internal combustion engine does not currently exist. It is clear that electric vehicles will be nothing more than a niche market, despite all the hype. It is impossible to replace all the worlds' vehicles with electric powered equivalents; therefore ensuring you can power your diesel vehicle should be a priority to anyone in the developed world.

---

2 https://www.theguardian.com/politics/2017/jul/25/britain-to-ban-sale-of-all-diesel-and-petrol-cars-and-vans-from-2040- 24/02/2019

However, despite the low cost of petrol/diesel fuels & thousands of gradual enhancements, developments to both the internal combustion engine & motor vehicles in general since 1861, the development focus has now shifted away from petrochemical powered vehicles. Most governments therefore appear to be moving towards electric vehicles as a replacement for petrol/diesel vehicles, but these are not a suitable or viable alternative. Until a credible replacement is found & any associated infrastructure built to support it, the majority of the vehicles on the road will remain to be either petrol or diesel powered. It will therefore be advantageous to understand exactly how a diesel engine works & how diesel is made to understand how the engine can be powered without diesel fuel, but with biodiesel.

The Diesel engine

As was stated previously, the Diesel engine was conceived by Rudolph Diesel, who in 1892 was granted a patient for his New Rational Combustion Engine. This was a compression-ignition engine (CI-engine), where the air in the cylinder is compressed, which results in the compressed air becoming heated & in turn this heat spontaneously combusts the fuel (this is known as adiabatic compression). If you have pumped up a bicycle tyre & felt the pump getting hot, this is because the compressed air is becoming hotter.

There is no spark ignition which is seen in petrol engines, it is a purely mechanical process which causes the combustion of the air-fuel mixture. This combustion then rapidly expands, pushing the cylinder downwards. Diesel's early engines used various fuel sources. His first design used coal dust as a fuel source. By 1900, he exhibited an engine that ran successfully on peanut oil. His early designs would allow the engine to run on virtually any fuel, including vegetable oils. Diesels concept was to provide an engine that people could fuel with whatever fuels or oil's they had to hand.

Figure 2 Diesels first experimental engine 1893 (Imotorhead64 © 2008) from 'Rudolf Diesel: Die Entstehung des Dieselmotors. Springer, Berlin 1913. ISBN 978-3-642-64940-0. p. 11, Fig 3'

Rudolph Diesel died in September 1913 in what can only be described as mysterious circumstances. It is not clear if he committed suicide or if he was murdered.

Following his death, his CI-engine was then further developed by various companies. His engine (which has now simply become known as a diesel engine) was then used in agricultural machinery, stationary engines, submarines & ships. Further developments over the years reduced its size allowing it to be used in trains & trucks.

More enhancements were to follow over further years, reducing the engine's size even further, allowing it to be small enough to be used in vans & cars. Currently, all of these diesel engines use an oil based fuel, which is also eponymously known as diesel, but there is no reason why the diesel engine cannot still use other fuels.

The production of diesel fuelled engines in cars started in 1933 when Citroën featured a diesel engine in their Rosalie.

Figure 3 Citroën Rosalie (Arnaud 25 © 2009)

1933 also saw the first diesel fuelled car enter the Monte Carlo Rally, which was a 1925 Bentley fitted with a Gardner 4LW engine & driven by Lord Howard de Clifford. It finished as the top British entry & finished 5th overall.

Following several diesel powered Mercedes-Benz taxi's which were not popular, only a small number of other vehicles had a diesel powered option at that time. They were the Austin A60 Cambridge, Isuzu Bellel, Fiat 1400-A, Standard Vanguard, Borgward Hansa & a few others.

Diesel cars were never a popular choice until 1982 when Peugeot Citroën introduced their XUD engine in the Peugeot 305, Peugeot 205 & the Talbot Horizon. This XUD engine remained a world leading diesel engine until the mid 1990's.

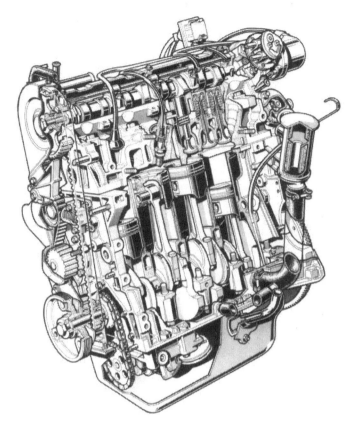

Figure 4 Peugeot Citroën XUD engine (Public Domain © 2020)

The first mass market turbo diesel was found in the XUD powered Citroën BX (1988) & the Peugeot 405 (1989). These cars & engines proved to be so good, they paved the way for diesel fuelled cars to achieve an almost 50% market share in the EEC today.

The XUD engine is indirect injection (IDI) & was the forerunner of the current diesel engines. Interestingly, because the XUD engine was IDI, the combustion burn time was slow, therefore it is suitable to run straight vegetable oil (SVO) without any modifications.

The Fiat Croma TD-i.d. was the first to have turbo direct injection in 1987. This had mechanically controlled injection, but this was then bettered with the Audi 100 TDI in 1989 as they included electronic control of the direct injection system.

The Alfa Romeo 156 (1997) was the first common rail diesel car & in 1998, a diesel powered car won the 24 Hour Nürburgring.

In 2004 Honda produced their first diesel engine (i-CTDI) & fitted it in their Honda Accord. In 2006 a diesel fuelled Audi R10 TDI LMP1 won 24 Hours Le Mans.

Diesel design

The modern diesel engine can be classed as an internal combustion engine, but differs from the petrol engine somewhat. Diesel engines are far simpler which means they are more robust & as a result tend to break down far less than a petrol fuelled engine.

Firstly, in the diesel engine, (as previously stated) it is air that is drawn into the cylinder, not an air-fuel mixture like in the petrol engine. The air in the cylinder is then compressed to between 14 to 25 times its original volume. A petrol fuelled engine will only compress the air-fuel mixture to about 10 times its original volume.

As a diesel engine has close tolerances to achieve the high compression ratio, the air entering the engine must be clean & free of any debris. It is therefore important to change the air intake filter at regular intervals. The air intake point is also recommended to be at a point as far away from the engine as possible, so that the intake air is cool. Cool air is more oxygen dense than hot air, so the cooler the air temperature, the better the compression in the cylinder.

Compressing any gas will always result in an increase in temperature. In the diesel engine, the compressed air needs to achieve a temperature which is a minimum of 500°C (932°F). There are glow plugs to aid this process, not spark plugs which are only found in petrol engines. The diesel fuel is then injected into the cylinder where it is ignited by the hot, compressed air. The resulting explosion then pushes the piston downwards just like it would in the petrol engine. When the piston returns back up the cylinder, it expels the hot gas through the exhaust valve just as in a petrol engine.

That piston is attached to the crankshaft by a rod & therefore that downward movement is what ultimately moves the crankshaft. The glow plugs are only present to help warm the air whilst the engine is cold.

This all operates in the same controlled sequence, in what is known as four strokes.

Stroke 1, **Intake**, air is drawn into the cylinder through the intake valve as the cylinder is drawn down.

Stroke 2, **Compression**, the intake valve closes & the piston moves upwards compressing the air & heating it to at least 500°C. A small amount of diesel fuel is injected into the cylinder where it spontaneously ignites & therefore explodes.

Stroke 3. **Power**, the resulting explosion pushes the piston back down again, thus turning the crankshaft through the con-rod which connects them both.

Stroke 4. **Exhaust**, the outlet valve then opens & as the crankshaft continues to turn, it pushes the piston back up again & this action pushes the hot gas that remains after the explosion, out of the cylinder then down the exhaust. The cycle then repeats itself again & again.

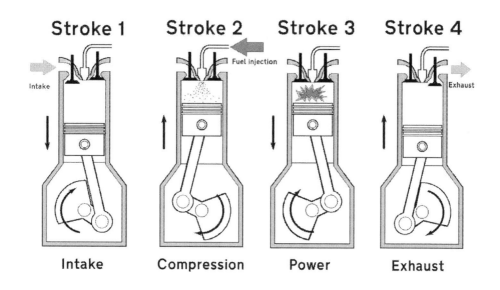

Figure 5 The Diesel 4 stroke cycle (P Xavier © 2020)

This is a similar four stroke sequence as a petrol engine. It has the same downside, which is, power is only provided on the 3rd stroke. Again, to overcome this issue, four cylinders are generally used together & in sequence. Therefore if cylinder 1, is on stroke 1, cylinder 2 is on stroke 2, cylinder 3 is on stroke 3 & cylinder 4 is on stroke 4. Power is therefore provided on every stroke from one of the cylinders & the result is no loss of power. Typically, a diesel engine has 2, 4 or 6 cylinders, but the most common configuration is 4. There are also some 2 stroke diesel engines, but the majority of diesel engines are 4 stroke.

Even though there is very little mechanical difference between the petrol & diesel engines, the diesel engine actually works out to be approximately 40% more efficient, therefore 40% more miles can be travelled in a diesel vehicle than a petrol vehicle using the same volume of fuel. The diesel engine achieves this due to numerous factors.

Firstly, Carnot's rule[3] states that the efficiency of an engine is dependant on the difference between the highest & lowest temperatures in which it operates, therefore if the diesel engine is very hot, or if the external air temperature is very low, then the difference between the two is high & therefore it is more efficient.

Secondly, because of the simplistic design of the diesel engine, it allows the fuel to achieve a higher temperature & therefore removes the need for any spark plugs. The higher temperature also results in a higher percentage of the fuel burning & as a result it releases a greater amount of energy. Therefore some turbo diesels use an intercooler to reduce the temperature of the air being fed into the engine.

Thirdly, as the design allows for a higher percentage of power output, the engine operates at a lower output ratio when compared to a petrol engine. Therefore the petrol engine will burn more fuel than a diesel engine to achieve a comparable power output.

Fourthly, as diesel fuel contains more energy than petrol, again less fuel is needed when compared to petrol.

Finally, diesel fuel also acts as a lubricant which allows a diesel engine to operate with less friction & this will result in a longer lifespan for the engine components & less overall maintenance. A diesel engine is not throttled like a petrol engine; therefore the amount of air drawn in at any engine speed is always constant. The engine speed is regulated by the amount of fuel injected into the cylinder. The more fuel injected into the cylinder, the larger the resultant explosion & therefore the more power is produced.

---

3   https://en.wikipedia.org/wiki/Carnot%27s_theorem_(thermodynamics)   —
      02/03/2019

The accelerator pedal is therefore connected to the metering unit of the engine injection system rather than to the air intake flap which is found in a petrol fuelled engine.

Diesel engines produce more power than a petrol equivalent, but operate more slowly. They are also heavier. They are therefore more suited for pulling power, not speed. This is why there are very little diesel powered racing cars. Manufacturers therefore tend to add turbochargers to modern diesel engines, in an attempt to boost their performance.

Diesel injection

All diesel engines have a system of injecting fuel into the cylinder. Some are direct injection (injecting directly into the cylinder, others are indirect (injecting into a premixing chamber). There are also various small adaptations between the systems with different manufacturers.

The injector mechanism is how fuel is directed from the tank to the injector. There are three types, a rotary fuel system, an individual fuel system & the common rail system. The difference is in the high pressure pump used.

The rotary fuel system uses a distributor or rotary vane pump. The system has one shaft with a single plunger. Despite there being four injectors, the number of plungers remains single & is located within the pump shaft which is rotating. At specific points during every rotation, there are fuel barrels, which when the plunger passes through them, the fuel will be injected into one of the injectors. Therefore, if there are four cylinders, there will be four injectors & therefore four fuel barrels which surround the pump shaft.

The Individual fuel systems use a pump of the individual inline type. Where, the number of plungers will be adjusted by the number of injectors present.

This is because each plunger will serve just one injector, so that if there are four injectors, there will be four plungers which will all be arranged in line. The system operates from the camshaft as on each pass, the plunger will be pressed giving the right timing. When the plunger is pressed against the cam, the fuel will be sprayed into the cylinder. Therefore the number of cams will be equal to the number of injectors & that the cam angle is will be adjusted according to the ignition timing.

Since the 1960's, diesel engines of the common rail type used a low pressure fuel pump to feed fuel into the injectors or pump nozzles; this was later changed to a high pressure pump. It is more typical now to use two pumps. The first pump is to transfer the diesel from the fuel tank to the fuel line, then the second pump is high pressure, which increases the fuel pressure. If only one pump is used in the system, it is a high pressure pump. The pump used on a common rail system is always continuous; therefore the pump will continue to press fuel with a stable pressure.

The latest development came in the 1990's when Denso invented the piezoelectric injector which offers a greater level of precision. These injectors are controlled by the vehicle's engine control unit (ECU). The whole system allows for the ECU to make fine adjustments to the fuel injection timing, quantity & pressure which all allows for better atomisation of the fuel. As such, these engines have a very short heating up time, make less noise & have lower emissions when compared to the older designs.

All direct fuel injection systems cost more than indirect, as the injectors are exposed to far more heat & pressure, therefore more costly materials are used that are machined to higher tolerances.

Indirect systems are older. They rely on far simpler injectors which are injected into a sub chamber which is shaped to swirl & mix the compressed air & diesel mixture to achieve better combustion.

It is less efficient than the now common direct injection system. With the older indirect injection systems, because the fuel is injected into a sub chamber the air is warmed by the engine. This makes the whole engine 5 – 10% less efficient than a direct injection system.

Current diesel engines use either a 'common rail' system (using rail fed solenoid valves to feed the unit injectors), or a 'unit injector' system (which has the injector nozzle & injector pump housed in a single unit). Older common rail systems tend to be solenoid design, whilst newer systems use piezoelectric actuators.

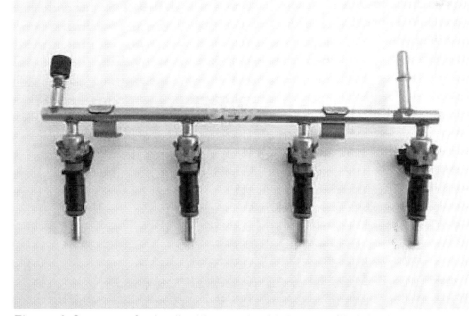

Figure 6 Common fuel rail with attached injectors (Public Domain © 2020)

In either common rail systems, each injector is electronically controlled allowing the optimal delivery of fuel for injection into the chamber. Also, the common rail eliminates the need for high pressure hoses within the fuel delivery system. There is also a system that combines direct & indirect systems called 'helix controlled direct injection'. This system employs a combustion cup in the head of the piston into which the diesel is sprayed.

The majority of these newer injection systems are designed to provide huge injection pressures. Many of them can be in the range of 7 to 70 mega-pascals (1,000 to 10,000 psi). Although each & every system listed here sprays the diesel as an aerosol, into the combustion chamber just before the cylinder is at top dead centre (TDC), which is just before maximum compression is reached.

Turbo charging & super charging

As was previously stated, air in a diesel engine is compressed in the cylinder. The turbocharger is used to compress more air into the engine cylinder by compressing it (to above atmospheric pressure) before it reaches the combustion chamber. The simple act of compressing the air before hand increases the overall power output of the diesel engine.

Figure 7 A diesel turbocharger (Public Domain © 2020)

A turbocharger is made from two main components. The first part is the turbine wheel within its housing. The second part is the compressor wheel inside its housing. The exhaust gases are fed into the unit where the high velocity turbine wheel spins. This spinning draws in the air & compresses it.

The compressed air is then channelled into the compressor where it is converted to a high velocity air flow by a process called diffusion. This high velocity, compressed air is then channelled to the air intake of the engine which allows it to burn more fuel & therefore produce more power.

A supercharger works in much the same way as the turbocharger, insofar as it compresses air (to above atmospheric pressure) & channels it into the engine which in turn allows it to burn more fuel & therefore produce more power. However, where the turbocharger was driven by the exhaust gasses, a supercharger is powered by the engine, typically through a belt, gear, chain or shaft that is connected to the engine's crankshaft. Due to the fact that they receive their power from the crankshaft, they will therefore reduce the overall power that the engine has produced. They can reduce the overall output by 33%. It is for this reason that the majority of diesel powered vehicles on the road are turbocharged, not supercharged. However, there are a still a small number of new supercharged diesel vehicles being made today.

There are three types of superchargers. The first of which is the 'Roots type supercharger'. This uses rotors that blow air into the intake port. As the port contains more air, it becomes compressed. Another type is the 'twin-screw supercharger'. This compresses the air within the supercharger before it is channelled to the intake port. It does this by employing two screw type rotors that draw the air in & then compresses it. The third & final type is the 'centrifugal supercharger'. This uses an impeller fan to suck in the air.

Many of the turbocharged & supercharged engines also use an intercooler to cool the air that has been compressed. It is in essence just like a small car radiator that the air exiting from the turbo or supercharger will pass through or over, which will then reduce the temperature of the air. This reduction in air temperature will then further increase the density of the air which is routed back to the engine air intake.

In addition, some manufacturers have even tried twincharging. That is by fitting both a turbocharger & a supercharger to the same engine. In the 1985 & 1986 World Rally Championships, Lancia fitted a twincharger to their Lancia Delta S4, only to find the increased complexity impacted negatively on the cars reliability during the races. Volkswagen also has a TSI engine that uses a twincharger, as does the Volvo S60, XC60, S90 & the XC90.

Diesel particulate filter & catalytic converters

A diesel particulate filter (DPF) is designed to remove diesel particulate matter (soot) from the engine exhaust gasses before it exits the vehicle. The DPF therefore acts as a filter to collect the soot. It differs from a catalytic converter (CAT) insofar as the CAT changes the chemical composition of the exhaust gas rather than filtering it. The exhaust gasses passes through the CAT where a reaction takes place on the surface of a ceramic block that is coated with a mixture of palladium, platinum & aluminium oxide. These catalytically oxidise the hydrocarbons & carbon monoxide molecules with oxygen to form carbon dioxide & water molecules. The DPF is used solely with diesel engines but the CAT can be found paired with any internal combustion engine.

DPF's can remove 100% of the soot from the exhaust gasses, but in real world conditions typically operate at 85% efficiency. Some are designed to be replaced at certain intervals, such as when the vehicle is serviced, & some are designed to burn off the accumulated soot either by using a catalyst or by heating the DPF to achieve soot combustion temperatures.

Many have a program built into the vehicles electronic control unit (ECU) that will initiate a burn when the DPF requires it. Any cleaning of the DPF is known as 'filter regeneration'.

This is achieved at motorway speeds rather than urban speeds, therefore it is advisable to drive a diesel with a DPF on a motorway occasionally & immediately if a warning light appears on the vehicle dashboard. Some newer diesels can perform a 'parked regeneration', where the vehicle is parked up & the engine increases the RPM to a high level whilst parked.

The fuel injectors heavily influence the size & formation of the soot particles as they are formed from incomplete combustion of the diesel fuel. A variety of other particles are produced during combustion of the fuel - air mix due to incomplete combustion. The composition of these particles will vary hugely & it is dependent upon the engine type, its age & the level of emission specification that the engine was designed to meet. The particles generated are smaller than a one micrometer (one micron) in size. Also, high sulphur diesel fuel produces far more soot than low sulphur diesel fuel.

There are several types of DPF's currently on the market. They are, cordierite wall flow filters, silicon carbide wall flow filters, ceramic fibre filters, metal fibre flow filters & partial filters.

Wall flow filters are mostly used by vehicle manufacturers & both cordierite & silicon carbide DPF's are wall flow filters. They look like ceramic cylinders that have small holes drilled through their length. Every other hole is blocked at each end of the cylinder. This forces the exhaust gasses down the open channels, then through the channel walls into the neighbouring channels, where they are free to exit from the rear of the DPF.

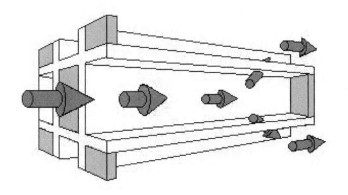

Figure 8 Exhaust gas flow (P Xavier © 2020)

Cordierite wall flow filters are the most common in use. They are made from cordierite which is a ceramic material that is also used in CAT's. They provide an excellent level of filtration efficiency, are inexpensive to produce, but as cordierite has a melting point of approximately 1,200°C (2,192°F), they have been known to melt during regeneration. However this can only happen if the DPF contains far more soot deposits than it is designed to hold.

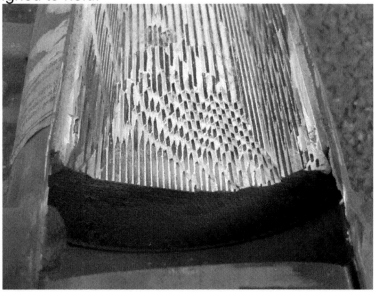

Figure 9 Cutaway of a blocked DPF (P Xavier © 2020)

The second most common type of DPF is the silicon carbide wall flow filter type. Silicon carbide has a higher melting point (approximately 2,700°C – 4,892°F) than cordierite, but is more difficult to produce into a useable product. Silicon carbide DPF's are therefore far more expensive than their cordierite equivalents.

Ceramic fibre filters are manufactured from several different types of ceramic fibres which are mixed to form a porous fibrous material. The density of the material can be controlled during manufacturing which allows for the filter to be designed to cater for high flow - lower efficiency or high efficiency - low volume filtration. These filters achieve efficiency levels greater than 99%, but need regular regeneration or the filter will quickly become blocked.

Metal fibre filters are very similar, except they are made from metal fibres. These have the advantage of allowing an electrical current to pass through the filter, which allows the filter to be regenerated at a low temperature. These metal filters therefore tend to be the most expensive due to the integrated electrics & associated electrical wiring & control systems on the vehicle.

Partial filters can be made from a variety of different materials & can achieve a filtration level between 50 – 85%. This type of DPF is generally used for the retro-fit market for older diesel vehicles that have no factory fitted DPF.

The diesel particulate matter will naturally burn off at 600°C (1,112°F), but this temperature can be reduced to between 350 – 450°C (662 – 842°F) if a chemical additive that is designed to clean the DPF is added to the diesel fuel.

As previously stated, a catalytic converter (CAT) is completely different to a DPF as it changes the chemical composition of the exhaust gas rather than filtering it. These are known as a 'diesel oxidation catalyst' (DOC). The DOC converts the carbon monoxide (CO) into carbon dioxide (CO_2) & also breaks down any un-burnt fuel.

The DOC is typically the smallest component of a diesel CAT & is also the one with the highest monetary value with regard to recycling. This is because the majority of the platinum & palladium is located in this part of the CAT. It is also almost always the first component within the exhaust system.

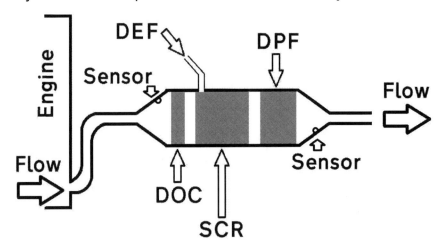

Figure 10 Typical diesel CAT layout (P Xavier © 2020)

The next section is known as the 'selection catalytic reduction' (SCR). This section is usually the largest, but has the lowest monetary value when it comes to recycling. This is the section which has the task of reducing the nitrogen oxide (NOx – a composition of nitric oxide (NO) & nitrogen dioxide (NO_2)). In a three-way CAT, rhodium is the precious metal that catalyzes the reduction reaction, which is when the NOx is converted into nitrogen (N_2) & oxygen (O_2). However, because of the very high level of oxygen in the diesel exhaust, rhodium (Rh) cannot efficiently reduce the nitrogen oxide (NOx). Therefore, diesel converters typically do not contain any rhodium (Rh).

Due to the ineffectiveness of rhodium (Rh), the SCR uses an additive called 'diesel exhaust fluid' (DEF). This is usually 32.5% urea (CH_4N_2O) & 67.5% deionised water. It is sold under the name AdBlue. Sometimes it is called AUS32, which means aqueous urea solution 32%. It is added to the SCR at a ratio of between 2 – 6% of the diesel consumption. The exact ratio is controlled by the vehicle's ECU & is dependant on various parameters such as temperature & speed.

The AdBlue is kept in a separate tank to the fuel & whilst the engine is running, tiny amounts of the AdBlue is squirted onto the SCR & as the exhaust gas passes through the SCR, the AdBlue converts the nitrogen oxide (NOx) into nitrogen (N_2) & water (H_2O). The position in the exhaust for the DPF & the SCR is not important, so long as they are placed down stream from the DOC.

The SCR is an important element in the catalytic converter as nitrogen dioxide (NO_2) is the gas that causes the most concern with inner city pollution levels because it is harmful if inhaled in large concentrations. However, it is not just diesel engines that produce nitrogen dioxide (NO_2). It will be produced in any high temperature combustion scenario, so this will also include petrol engines & even gas powered central heating systems.

Lambda sensor

At the start of the exhaust system, before the DOC there is a sensor. This is called an oxygen sensor or lambda (λ) sensor. This is used to measure the proportion of oxygen (O) in the exhaust gasses. There is also typically another oxygen sensor after the DPF. Both sensors monitor the oxygen level in the exhaust so that the ECU can determine & if required adjust the air-fuel ratio in the combustion cylinder whilst ensuring the whole exhaust system is operating within specified limits, whether the fuel to air mix is optimal & whether the exhaust system needs regeneration or not. These allow the ECU to fine tune the combustion & exhaust to reduce pollution.

These were the main elements found in a diesel engine vehicle that are diesel specific, or greatly enhance the diesel engine. If a simpler or more detailed explanation is required, there are numerous internet sites that give excellent explanations. Many of these sites also use diagrams & videos. A simple internet search will reveal thousands of sites in various languages.

Chapter 2 – Diesel fuel

All the worlds' diesel fuel including that which you use in your vehicle is all derived from the same source. That is crude oil. It is crude oil that is pumped out of the ground & this is what the fossil fuels that are used in the world's vehicles are all made from, therefore it's important to understand how it's processed.

Crude oil

The crude is sold by the barrel, which is a barrel containing 159 litres (because the barrel unit was standardised in the 1870's & based on a whiskey barrel). The crude oil is then refined & separated (by a process called distillation[4]) into its residual components. On average between 40 – 70 litres of petrol can be separated per barrel (but it is dependant on the actual make up of the crude oil, where the crude oil originated & the process used to refine it). OPEC state that 70 million barrels of oil are produced every day, which equates to almost 49,000 barrels (7,791,000 litres) every minute[5].

There are around 1,500 products that can be made from the crude oil in this way, but the most common are listed here (annual percentage from crude oil). 44.1% petrol (aka gasoline), 20.8% distillate fuel oil such as diesel oil & heating oil, 9.3% kerosene type fuel oil, 5.2% residual fuel oil, 4.3% liquefied refinery gasses, 4.3% still gas, 4.1% coke (the fuel), 2.9% asphalt & road oil, 2.7% petrochemical feedstock's, 1.1% lubricants, 0.5% kerosene & 0.7% other items.

---

[4]

https://en.wikipedia.org/wiki/Continuous_distillation#Continuous_distillation_of_crude_oil – 09/03/2019

[5] https://www.nationalgeographic.org/encyclopedia/petroleum/ - 10/03/2019

The crude oil naturally forms underground in reservoirs. When these are tapped to remove the contents, it is called a well. There is more pressure underground & also more heat than is found at the surface & as numerous other elements are dissolved in with the crude oil, when it is pumped to the surface, some of those dissolved elements can then begin to escape. For instance flammable gasses can escape from the crude oil as they are no longer confined at a high pressure. Hydrocarbons can also escape as solids. The exact chemical makeup of each of these underground reservoirs will vary from well to well, so some produce a high percentage of gasses (these are gas wells); some will produce a high proportion of crude oil (oil wells). Also, the exact molecular composition & proportion of crude oil & gasses will vary from well to well, but the chemical elements in crude oil will always fall within the following ranges. 83-85% carbon, 10-14% hydrogen, 0.1-2% nitrogen, 0.05-1.5% oxygen, 0.05-6.0% sulphur & <0.1% dissolved metals.

Hydrocarbons

The level of hydrocarbons contained within the crude oil also varies from well to well, but they will always fall within the following ranges. 15 – 30% alkanes (paraffin), 30 – 60% cycloalkanes (naphthenes), 3 – 30% aromatics & the remainder being made from asphaltics.

Hydrocarbons are just carbon molecules with attached hydrogen molecules. In the crude oil these are found in chains of molecules, which are in the range of between 5 – 40 molecules in length. It is these chains that are used to make fuel. The chains of 1 – 4 molecules long form gasses, the chains being 5 – 8 molecules long are refined into petrol. Those being 9 – 16 molecules long are refined into diesel, kerosene & jet fuel. The ones over 16 are refined into fuel oil & lubricating oil. Whilst the chains of around 25 are made into paraffin wax. Asphalt is made from chains of molecules that are 35 & over.

This will be explained in further detail in the organic chemistry chapter.

Figure 11 demonstrates how these hydrocarbon chains are strung together. The first is a chain of 3, therefore it is a gas molecule. The second is a chain of 6, therefore it is a petrol molecule. The third is a chain of 11, therefore it could be used in diesel, kerosene or even jet fuel. The final is a chain of 17, therefore it could be used for either heating oil or lubricating oil.

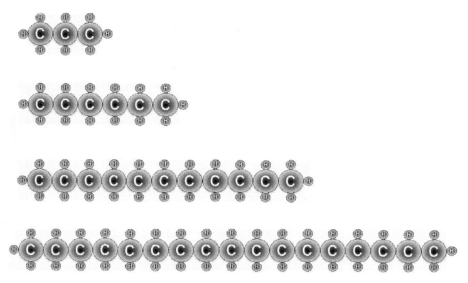

Figure 11 Carbon chains (P Xavier © 2020)

Fractional distillation

These chains of molecules are all obtained by boiling the crude oil at various temperatures at a refinery with a process known as fractional distillation[6]. It is called this as fractions are removed at varying temperatures.

---

[6] https://en.wikipedia.org/wiki/Fractional_distillation - 09/03/2019

As the temperature increases, the products are removed, all at controllable & predictable temperatures.

Liquefied petroleum gas (LPG) is obtained from the crude oil at a temperature of -40°C. Butane is extracted between -12 to -1°C. Petrol is liberated in the range of -1 to 110°C, whilst jet fuel is produced at 150 – 205°C. Kerosene is removed at 205 – 260°C & fuel oil between 205 to 290°C. Diesel is then taken out at in the temperature range between 260 to 315°C. Petrol, diesel & flammable gasses & all the other petrochemical fuels are therefore refined from the crude oil by the same fractional distillation process. As the temperature rises, the crude oil releases its contents at the specified temperatures.

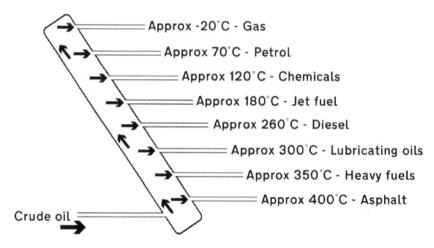

Approx -20°C - Gas
Approx 70°C - Petrol
Approx 120°C - Chemicals
Approx 180°C - Jet fuel
Approx 260°C - Diesel
Approx 300°C - Lubricating oils
Approx 350°C - Heavy fuels
Approx 400°C - Asphalt
Crude oil

Figure 12 Fractional distillation (P Xavier © 2020)

It should be noted from the previous diagram that it is cooler at the top & much hotter at the bottom. The products that are liberated from the crude oil at lower temperatures (at the top of the stack) have a low viscosity (runs easily as a fluid or gas) & are highly flammable (ignites & burns easily). Then at each incremental increase, as the temperature rises & as each product is released from the crude oil, the viscosity of the products will rise (they will become thicker & stickier) but the flammability will drop (will not ignite or burn easily & will burn with a sooty flame).

Therefore it can be clearly seen that petrol will be highly flammable & have a low viscosity, but when the asphalt is liberated from the crude oil, it will be far less flammable & also extremely viscous.

Flammability & viscosity

There is therefore a direct link between the length of the carbon chains to both flammability & viscosity. Also, with combustible fuels, such as petrol, the flammability is measurable. This is expressed as a research octane number (RON). The higher the octane number, the more compression the fuel can withstand before detonating (igniting). Fuels with a higher octane rating are used in high performance engines which require higher compression ratios. Petrol engines operate with the ignition of air & fuel which is compressed together as a mixture. It is this mixture which is ignited at the end of the compression stroke using the spark plugs.

Therefore, a high compression ratio of the fuel matters for petrol fuelled engines. If petrol with a low octane number is used, it may lead to the problem of engine knocking, which is when un-burnt portions of fuel is heated or compressed too much. Any un-burnt fuel can then self ignite before or after the desired time & as a result cause damage to the engine.

In the UK, the petrol available at the forecourts are usually rated at 95 RON for unleaded petrol. Many also have what they call 'super' which has a 97 RON rating, but this is only suitable for performance engines.

As a comparison, a small petrol powered two stroke outboard motor on a small boat would require a fuel with a 69 RON rating, with the highest rated freely available petrol having a 102 RON rating, but there are very few performance engines requiring a fuel of this level. Diesel oil has a rating somewhere between 15 – 25 RON. Within the EEC, 5% ethanol can also legally be added to the petrol.

Ethanol has a 108.6 RON. Liquefied petroleum gas (LPG) is a mixture of propane & butane, just propane or just butane. Propane has a 112 RON, whilst butane is 102 RON, therefore LPG is in the region of 102 – 112 RON, but it depends on the proportional mix of the gases.

Typically, diesel fuels do not have a RON rating. Instead they are given a cetane number. This is used as an indicator to the speed of combustion & also the compression needed for the ignition. The cetane number (CN) is important when determining the quality of diesel fuels. The higher the CN, the shorter the delay before ignition. Within the EEC, the current minimum standard is 51 CN for regular diesel, whilst premium diesel fuel can have a CN as high as 60.

Biodiesel is generally in the range of 46 – 52 CN & animal fat based biodiesel often falls within the range of 56 – 60 CN, whilst waste vegetable oil (WVO) typically has a value of approximately 48 CN.

Petrol has a limited shelf life. In a sealed container it will store for approximately one year. If the seal is broken, it will have a shelf life of just six months at 20°C & only 3 months at 30°C. If it is left open to the elements, given time, the petrol will also completely evaporate.

Also, as petrol contains various components & each of these components degrade or evaporate at different rates, over time, the composition of the petrol will change significantly from its original design specification.

Diesel fuel does not evaporate, but it will also store for one year within a sealed container. After this, it will start to turn to gum & the diesel will then tend to block filters in the engine. If diesel is left open to the elements it will degrade as it will act as a growing medium for fungus & bacteria which will degrade the diesel. Therefore treating stored diesel every six months will prolong its lifespan. Biodiesel has similar qualities to diesel & will degrade in the same way as diesel, but its lifespan is shorter.

Forecourt diesel

Since 12th October 2018 in the EEC, all diesel fuel (along with petrol & gaseous fuels) sold at the forecourts must have a standardised label on the pump which is universal throughout all EEC countries. These labels must also be placed on any new vehicle (registered after 12th October 2018) in the proximity of the fuel flap, in the vehicles manual & also displayed at the vehicle dealership. This has been implemented by the EEC in what they call '*Alternative Fuels Infrastructure Directive 2018*' *(EU Directive 2014/94/EU)*. It was conceived with the aim of reducing the incidents of people misfuelling their vehicles. It also helps the EEC reduce its overall CO_2 emissions & therefore meet its climate change targets. In the UK this labelling system is implemented under '*The Alternative Fuel Labelling and Greenhouse Gas Emissions (Miscellaneous Amendments) Regulations (2019)*' & came into effect on 1st September 2019.

 Diesel fuel is therefore now identified with a black square that has rounded corners & a white background. Inside the square is the letter 'B' which has a number after it.

Currently the number is 7. The 'B' stands for biodiesel & the number is a percentage. Therefore, the symbol shown here shows that the petrochemical diesel could contain 'up to' 7% biodiesel. It does not mean it actually contains any biodiesel at all, but it could. It could contain 5%, it could have 2% & it may even have no biodiesel in it at all. It is a maximum percentage indicator.

 If the square contains the letters XTL instead of a B & a number, then this indicates that it is synthetic diesel which could be made from natural gas or even vegetable oil, but not from crude oil.

In mainland Europe, it is even possible to find B10 (which could contain up to 10% biodiesel) & also B20 (which could have a maximum biodiesel content of 20%). This mixing of petrochemical diesel with biodiesel has been happening behind the scenes for the last 10 years. It is only since this labelling system has come into force that the whole concept of diluting petrochemical diesel with biodiesel has been widely publicised. Currently, the UK government's Department for Transport (DfT) now has a website to educate the public on this new labelling system[7].

The EEC is not the only place where this labelling system is employed. The EEA countries along with Macedonia, Serbia, Switzerland & Turkey are also adopting the EEC convention & in the USA, they have been using this numbering convention for some years, but there, it is even possible to see B100, which means it is not petrochemical diesel at all, but 100% biodiesel. In the future, B10 & B20 should also become widely available in the UK.

Also in the USA, in 2007 the Environmental Protection Agency (EPA) passed a law that diesel must meet certain specifications. Those specifications then determine which grade of fuel it is.

#1 grade diesel has less energy components & is more expensive than #2 grade diesel. However, #1 grade diesel rarely has problems in cold weather, which is completely the opposite of #2 grade diesel.

#2 grade diesel is the most readily available at the forecourts around the world. This grade of diesel holds the highest amount of energy components & lubricant properties in one mixture & also offers the best fuel performance.

---

[7] http://www.knowyourfuel.campaign.gov.uk/

Winterised diesel fuel is a combination of #1 & #2 diesels that, when blended together, holds a higher proportion of #1 grade diesel fuel than #2. This diesel is designed to be used during the winter months when it is too cold to use #2 diesel.

Low sulphur diesel (LSD) has been classified as diesel fuel that contains no more than 500 ppm (parts per million) sulphur content, while ultra low sulphur diesel (ULSD) must maintain no more than 15 ppm sulphur content. Since 2007, the EPA has ruled that all highway diesel fuel sold in the USA must now meet this ULSD specification.

The B/number system adopted by the EEC was first employed in the USA, where the number denotes the percentage of biodiesel contained within the fuel.

Chapter 3 – Organic chemistry made simple

To understand fully, what the hydrocarbon chains are, it is important to have a refresher on some basic chemistry that you no doubt studied at school.

Chemical Lego

Chemistry is just like Lego. It is like having building blocks that join together. Some of the parts fit together & make a solid connection; some things join together & have a wobbly connection. All things in this universe are made from these building blocks, including you. Organic chemistry is a branch of chemistry that uses carbon as the main building block. As you are aware, you are a living carbon based life form, therefore what is described in this chapter is exactly how your body was chemically made & how it operates. But carbon also bonds to non living elements too, so drugs, plastics & dyes are made in exactly the same way – chemically speaking.

In a nutshell, a single particle of matter is an atom. When an atom joins with one or more atoms it is called a molecule. Therefore if an oxygen atom joins with two hydrogen atoms, it makes a water molecule. Many atoms can be connected together to construct various elements. When they are stable they are in equilibrium, which is the chemical elements state of nirvana. Every atom is therefore looking for its own nirvana. This happens to be (for an atom) having electrons whizzing around its outer shell. Each of the atoms therefore has receptors to which the electrons will attach. Different atoms have different numbers of receptors.

Hydrogen has just 1 receptor; therefore it is looking to attach something to that 1 receptor. Oxygen has 2 receptors.

Therefore if an oxygen atom attaches to 2 hydrogen atoms (as previously stated), it makes a water molecule.

$$H_2O$$

That is how it is expressed – chemically, but it just means two hydrogen atoms & one oxygen atom. All the receptors are now filled, so this new (water) molecule is stable.

It is stable as it has no empty receptors, so it is not looking to join with any other atoms; it is now content to be just a water molecule.

Figure 13 A water molecule - H2O (P Xavier © 2020)

Other atoms have a varying number of receptors. Carbon has 4, as we have seen, oxygen has 2 & hydrogen has 1. Therefore carbon atoms can bond to 4 other atoms, oxygen can bond to two & hydrogen can bond to just one. Chemists call this valency, so when they say 'a carbon atom has a valency of 4', it just means that it has 4 receptors so can & will connect to 4 other atoms. Within their outer shells, carbon atoms contain four electrons. It is these electrons that are the receptors that can bond with other atoms. When an atom is bonded to another atom, the first atom shares an electron with the other atom. The second atom also shares an electron with the first carbon atom. It is these carbon-hydrogen molecules that are known as hydrocarbons & as they share one bond, they therefore belong to the alkane family.

Chemical chains

These hydrocarbons can join into chains (or even rings) of molecules, just as was seen earlier. These chains will then follow certain rules. One of those rules is that the longer the chain, the less chemically active it is.

So when crude oil is split into its various products, the short chained molecules that are liberated first (gasses) will be actively looking to interact & join with other molecules. As each of the products are released from the crude oil, these molecular chains increase in length & they become less & less reactive.

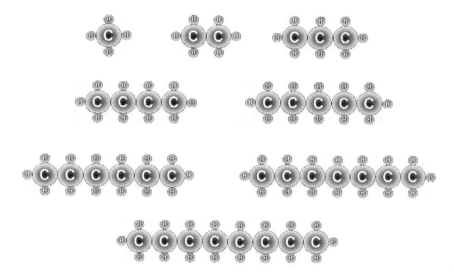

Figure 14 Hydrocarbon chains (P Xavier © 2020)

The above figure shows the chains as they increase in length. They step up, one molecule at a time as follows. CH_4 is methane (gas), C_2H_6 is ethane (gas), C_3H_8 is propane (gas), C_4H_{10} is butane (gas), C_5H_{12} is pentane (liquid), C_6H_{14} is hexane (liquid), C_7H_{16} is heptane (liquid) & C_8H_{18} is octane (liquid). Others in the sequence are C_9H_{20} is nonane (liquid), $C_{10}H_{22}$ is decane (liquid), $C_{11}H_{24}$ is undecane (liquid), $C_{12}H_{26}$ is dodecane (liquid), $C_{13}H_{28}$ is tridecane (liquid), $C_{14}H_{30}$ is tetradecane (liquid), $C_{15}H_{32}$ is pentadecane (liquid). $C_{16}H_{34}$ is hexadecane (liquid), $C_{17}H_{36}$ is heptadecane (solid), $C_{18}H_{38}$ is octadecane (solid), $C_{19}H_{40}$ is nonadecane (solid), $C_{20}H_{42}$ is isosane (solid), $C_{21}H_{44}$ is heneicosane (solid), $C_{22}H_{46}$ is docosane (solid), $C_{23}H_{48}$ is tricosane (solid), $C_{24}H_{50}$ is tetracosane (solid), $C_{25}H_{52}$ is pentacosane (solid), $C_{26}H_{54}$ is hexacosane (solid), $C_{27}H_{56}$ is heptacosane (solid), $C_{28}H_{58}$ is octacosane (solid).

$C_{29}H_{60}$ is nonacosane (solid), $C_{30}H_{62}$ is triacontane (solid), $C_{31}H_{64}$ is hentriacontane (solid), $C_{32}H_{66}$ is dotriacontane (solid), $C_{33}H_{68}$ is tritriacontane (solid), $C_{34}H_{70}$ is tetratriacontane (solid) & $C_{35}H_{72}$ is pentatriacontane (solid). There is no limit to the length of the chains that can be made, so the list can be limitless. By the time asphalt is liberated from the crude oil (made from a chain of more than 35 carbon atoms) it is stable enough to be used as a road surface as it actively repels water & oil. Also, the longer the chain, the more difficult it is to break the chain, this is because it is more stable.

Long carbon chains = low reactivity

The longer the chain of carbon atoms in the chain, the auto-ignition temperature will decrease, as can be seen in the following graph.

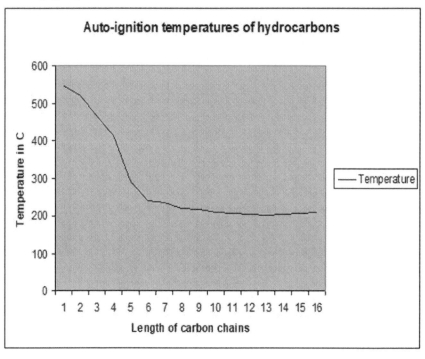

*Figure 15 Auto-ignition temperatures of hydrocarbons (P Xavier ©
2020)*

As previously mentioned, when an atoms receptor bonds with another atoms receptor so that they share one electron, the single bond makes the new molecule an alkane.

Chemical families

This can be thought of the family that it belongs to. Each of the examples seen so far have all been in the alkane family as they have shared only one bond. The alkane family is defined as 'acyclic branched or unbranched hydrocarbons that consist entirely of hydrogen atoms & saturated carbon atoms'.

But there are many other families. If these atoms share two bonds, they belong to the alkenes family. If three bonds are shared, then they belong to the alkynes family. Therefore, if ethane is used as an example, it can be seen that with a single bond the molecule is eth**ane** (alk**ane** family), if the same molecule shares two bonds it is eth**ene** (alk**ene** family), better known as ethylene or C_2H_4. If they share three bonds it is now eth**yne** (alk**yne** family), better known as acetylene or C_2H_2. These links are demonstrated in the following diagram.

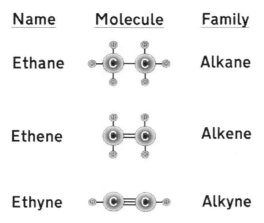

Name	Molecule	Family
Ethane		Alkane
Ethene		Alkene
Ethyne		Alkyne

Figure 16 Bonds (P Xavier © 2020)

Molecules in the alkyne group are quite unstable, as the bonds the atom shares with the other atom are only a quick fix for each of the carbon atoms. They would prefer to have another atom instead; therefore if two atoms (with a spare receptor each) bump into them, they will immediately form bonds with them & the molecule will then become an alkene. Again, that extra bond the alkene family has, is only a temporary fix, as if a further two atoms (with a spare receptor each) bump into them, they will immediately form bonds with them & the molecule will now become an alkane, which is nirvana for the molecule as it does not need to make any new bonds. It is now stable. The principles of these bonds & chains are the fundamentals of organic chemistry. But groups must also now be considered too.

Chemical groups & chemical bonds

If certain atoms join the molecule that will affect the chemical characteristic or reaction of the molecule, then it will belong in a specific group. The first of these groups are hydrocarbons; all the examples we have seen in the examples so far are hydrocarbons as they contain carbon, but there several others. Haloalkanes have a carbon – halogen bond.

There are numerous groups that contain a carbon – oxygen bond; there are carbon – oxygen groups that also contain nitrogen. There are groups that contain sulphur; groups that contain phosphorus; groups that contain boron & groups that contain metal. There are 67 known groups in total.

One of the groups that will be of relevance here is the alcohol group. That is an organic molecule that contains at least one atom from the hydroxyl group (oxygen bonded with hydrogen) which is also bonded to a carbon atom. Therefore, if an oxygen atom that is bonded with hydrogen is added to the flammable, colourless, odourless gas methane (CH_4), the result would be a light volatile, colourless & flammable liquid called methanol (CH_3OH).

If the same were applied to the colourless, odourless gas ethane (C_2H_6), the result would be the volatile, flammable, colourless liquid called ethanol (C_2H_6O).

If it was applied to the flammable, colourless, odourless gas called propane (C_3H_8), a colourless liquid called propanol (C_3H_8O) would be the result. It should be clear that just by adding one oxygen atom, the molecules properties have changed radically. Each of these three examples have now moved from being gasses to liquids & also all three now belong to the alcohol group, but all three still remain in the alkane family as they all still contain single bonds, as was first seen in figure 16.

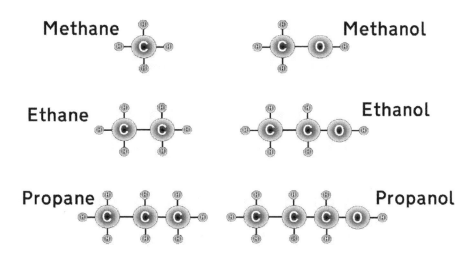

Figure 17 Simple alcohol molecules (P Xavier © 2020)

When the carbon atom shares two bonds with the oxygen atom it belongs to the carbonyl group. Again the chemical structure of the molecules can be radically changed. This time into either acids or esters which will move the molecules into the alkene family as they will all now contain double bonds.

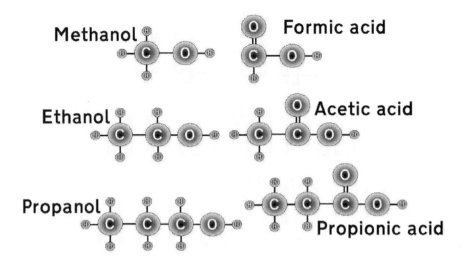

Figure 18 Simple acid molecules (P Xavier © 2020)

With the inclusion of a double bond, the light volatile, colourless & flammable liquid called methanol (CH_3OH) has been transformed into formic acid (CH_2O_2), which is a colourless liquid that has a pungent odour. In nature, formic acid can be found in most ants & they use it for defence purposes. The volatile, flammable, colourless liquid called ethanol (C_2H_6O) can be transformed into a liquid with an unpleasant smell. This is called acetic acid ($C_2H_4O_2$), which is the main constituent of vinegar. Using the same method, the colourless liquid called propanol (C_3H_8O) would transform into a liquid with a pungent, unpleasant smell similar to body odour. It is called propionic acid ($C_3H_6O_2$), as can be seen in figure 18. This is how a bottled wine (alcohol) can change into vinegar (acid) over time.

Esters can be made from an acid in which at least one hydroxyl molecule is replaced with a molecule from the alkyl group. Formic acid (CH_2O_2) can therefore be transformed into the most simple of the esters. The result of which is a highly flammable, colourless liquid with a delicate odour called methyl methanoate ($C_2H_4O_2$) aka methyl formate, which is a methyl ester.

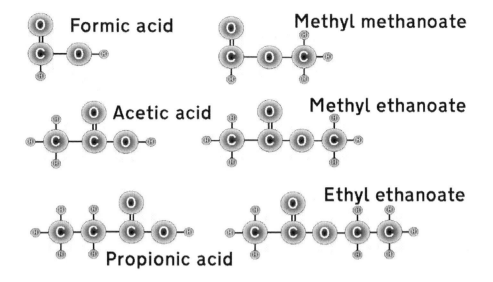

Figure 19 Simple ester molecules (P Xavier © 2020)

Acetic acid ($C_2H_4O_2$) can be changed into a flammable liquid which has a fruity odour, similar to glue or nail varnish remover. It is then called methyl ethanoate ($C_3H_6O_2$) aka methyl acetate. This is also a methyl ester. The pungent, unpleasant body odour smelling molecule called propionic acid ($C_3H_6O_2$) can be transformed in to a colourless liquid with an odour similar to pear drops. This molecule is called ethyl ethanoate ($C_4H_8O_2$) or ethyl acetate.

Employing these chemistry tricks to liberate or transform is exactly how biodiesel is made & just how to do this to make biodiesel will be looked at in a later chapter.

Chapter 4 – Sources used for fatty acid methyl esters (FAME)

There are many types of oils used for SVO fuelled diesel engines & used for the production of biodiesel. They are either derived from plant or animal origin. These oils & fats are commonly known as fatty acid methyl esters (FAME's).

First, some terminology must be known.

Saturated fats are fatty acids without double bonds. The lower the saturated fat content, the lower the gel point for any oil made from it.

Monounsaturated fats contain only one double bond in the molecular chain.

Polyunsaturated fats are when the chain contains two or more double bonds within the molecular chain. All fatty acids that contain double bonds are called unsaturated fats. Unsaturated fats have a higher gel point, but are more unstable & the bonds tend to break down more easily.

Hydrogenated oils or trans fats have been chemically altered to remove the double bonding. The purpose behind hydrogenation is to lengthen the shelf life of cooking oil, but it is then difficult to make fuels from hydrogenated oils as the whole process makes the oil more solid.

Oils that contain triple bonds are unsuitable for oils for SVO fuel or biodiesel use.

Unsaturated fats are more commonly found in plant & fish based material, whilst saturated fats are more commonly found in meat material & tend to be solid at room temperature.

First, to look at some plants that are suitable for oil production.

Plant derived FAME's

There are thousands of plant species that can produce oil which can be used to run a diesel engine. Therefore, ten have been chosen to be used here as an example. These ten have been chosen because their oils can be commonly found throughout the world. Not all can be grown universally, but wherever you live, there will be a few that can be grown in your region.

Arachis hypogaea – peanut

This is a legume that produces edible seeds, from which peanut oil can be produced. This plant is grown in both the tropics & the subtropics but is native to South America. It is a crop grown extensively in the southern states of the USA. In 2016, 44 million tonnes of shelled peanuts were produced globally. It is estimated that 3,700 litres of peanut oil can be produced per acre as the nuts contain over 50% oil.

This was the same oil that In 1900 Rudolph Diesel used to power his Rational Combustion Engine when he exhibited it at the World's Fair in Paris. Also, in the USA during WWII, peanut oil was used as a replacement for fossil fuel derived oils & lubricants when petrochemical derived oils were in short supply.

Peanuts have a high saturated fat content. This raises the gel point of the oil.

Figure 20 Arachis hypogaea (Public Domain© 1887)

Brassica napus – rapeseed

This is a bright yellow flowering plant of the Brassicaceae family (mustard or cabbage family).It is grown predominantly for its oil rich seeds & is thought to be one of the earliest plants to be cultivated by mankind, possibly 10,000 years ago.

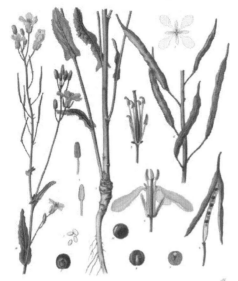

The seeds contain a high amount of erucic acid, which is toxic. A variety called canola is now grown as it was bred to be non-toxic.

It can be grown in more temperate regions such as Northern Europe. Since the EEC published their *Transport Biofuels Directive* in 2003, it has also been widely grown within the EEC.

Figure 21 Brassica napus (Public Domain © 1887)

In 2000, it was the world's third largest source of vegetable oil & 4,160 litres of oil can be produced per acre. It is one of the lowest saturated fats of any oil which makes it particularly suitable for use as an engine fuel.

There was a six fold increase in production between 1975 – 2007. Now it is estimated that 58.4 million tonnes of rapeseed was produced in the 2010 – 2011 growing season. The production of this crop is predicted to rise continually for the foreseeable future within both the EEC & USA due to its use as a cooking oil & its use for biodiesel.

Currently in the EEC, 80% of the EEC's rapeseed produce is used in the production of biodiesel. In the USA this plant & its oil is known as canola.

Cannabis sativa – hemp

The seeds from the hemp plant contain approximately 30% oil & it is for this reason & for its fibre, that it has been cultivated for at least 5,000 years in some parts of Asia.

In one archaeological dig site near Japan, hemp seeds have been dated to be over 8,000 years old.

Traditionally it was grown & used extensively to make rope & fabrics in Europe & North America until the beginning of the 20th Century. It has been illegal to grow hemp in the EEC since 1971 & the *'marihuana tax act'* 1937 in the USA killed off production there.

Figure 22 Cannabis sativa (Public Domain © 1887)

Henry Ford experimented with using hemp as a material to build a concept car in 1941. It was called the 'soybean car' & made from plastics & fibres all derived from wheat, hemp, flax & ramie. The concept was never developed due to the onset of WWII. But some vehicle manufacturers are now starting to use hemp in their vehicles. These include Audi, BMW, Ford, GM, Chrysler, Honda, Iveco, Lotus, Mercedes, Mitsubishi, Porche, VW & Volvo[8].

Oil from the hemp seeds contains 75 – 80% polyunsaturated fat & up to 11% saturated fat. It is therefore reported to have the highest concentration of unsaturated oil that can be obtained from any plant seed.

---

[8] https://en.wikipedia.org/wiki/Hemp - 22/06/2020

Carthamus tinctorius – safflower

This is another plant that has been found in the archaeological record to have been cultivated by humans. Textiles found in the tomb of Tutankhamen were even found to be coloured with dyes from the safflower.

It is estimated that in 2018, 627,653 tonnes of safflower seeds were harvested globally. Although the plant has many uses, over the past 50 years, it has been grown mainly for the oil that is derived from the seeds. More recently, a variety has been developed that can produce seeds that have a 93% oil content. The oil has also been found to be far superior than petrochemical oil when used as an engine lubricant.

Safflower oil contains saturated fat, monounsaturated fat & also polyunsaturated fats.

Figure 23 Carthamus tinctorius (Public Domain © 1887)

Currently over 60 countries grow safflower. Turkey is a major producer & is currently developing refining techniques to turn the oil into biodiesel. Australia is also working to develop the oil into a replacement for petrochemical diesel. In the USA, Montana State University's Advanced Fuel Centre have concluded that safflower lubricant oil under heat & pressure in a diesel engine, 'is better than what we see in petroleum'.

Cocos nucifera – coconut

The coconut tree is a member of the palm tree family & the coconut is not a nut, but a drupe, which is a fruit. The oil gels at around 10°C (50°F), so it is better suited to hotter climates.

The coconut is also believed to be another plant that has been cultivated by humans for thousands of years. This tree is to be found everywhere in the tropics & is believed to have been spread by Polynesian seafarers as they travelled across the Pacific & colonised the various islands.

In 2018, 62 million tonnes of coconuts were produced globally. Coconut oil has very high levels of saturated fat (88.5%).

Figure 24 Cocos nucifera (Public Domain © 1887)

The Philippines, Vanuatu, Samoa & a few other tropical island countries currently use coconut oil as an alternative to diesel fuel to run cars, trucks, buses & power generators. Some South Pacific islands are also undertaking studies into the feasibility of using the oil to generate their electricity.

The EEC may be set to replace the use of palm oil (within the EEC) with coconut oil as a study[9] showed that palm oil derived biofuels have three times the carbon footprint of petroleum diesel, which is all due to the deforestation of tropical forests & the subsequent planting of palm oil for commercial gain.

9

https://ec.europa.eu/energy/sites/ener/files/documents/Final%20Report_GLOBIOM_publication.pdf – 22/06/2020

Elaeis guineensis – oil palm

This species of palm is native to West Africa & is believed to have been cultivated over the last 5,000 years. Archaeologists have discovered palm oil in a tomb in Abydos, Egypt that dates to 3,000 BC.

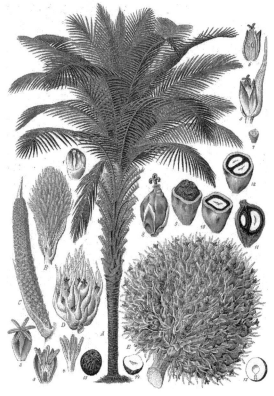

This species of palm will grow anywhere in the tropics within 20° of the equator & has now been naturalised in Cambodia, Central America, Indonesia, Madagascar, Malaysia, Sri Lanka & West Indies. It can also be found on several islands in the Indian Ocean & also on some Pacific islands.

It is so productive, an acre can produce nine times more oil than an acre of soya. In 2019, global production reached 75.7 million tonnes.

Figure 25 Elaeis guineensis (Public Domain © 1887)

Not only the oil can be harvested, some companies are looking into using the empty fruit branches & the palm kernel shells to generate electricity.

Despite the tree having natural credentials, a lot of deforestation has occurred when forests have been cleared to make way for palm oil plantations. Also, the oil produced fares worse when compared to fossil fuels as litre for litre, more carbon is released by this tree during its lifecycle than would be released by the equivalent amount of petrochemical fuel.

Glycine max – soybean

This has been cultivated in East Asia before records began. It is believed that the bean was domesticated in China around 7,000 BC & has remained a staple there ever since. Soy then became an important source of protein food in Europe & North America during WWII.

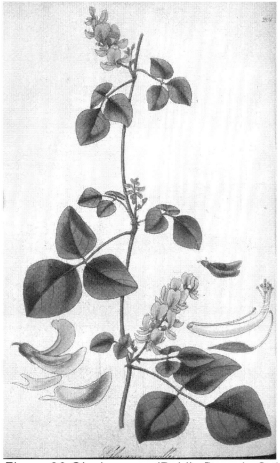

It currently provides 60% of the protein fed to livestock in Europe (who import 39 million tonnes per year) & in the USA 80% of their soy output is used in the production of biodiesel. In 2017 the global production was 352.6 million tonnes which was grown on 123.6 million hectares.

An acre of soybean can produce up to 2.8 tonnes per year. 300 million hectares of tropical forests been deforested over the past 20 years to make way for soybean plantations.

Figure 26 Glycine max (Public Domain © 1804)

In 2016, 18% of global soybean production was used to produce biodiesel, but the worlds major growers are USA (35%), Brazil (29%) & Argentina (18%), but because these soybeans need to be shipped to where they are needed to produce biodiesel, the soybeans have a high carbon footprint.

Helianthus annus – sunflower

This plant is believed to have been domesticated in Mexico around 5,000 years ago. It was brought from the Americas to Europe in the 16th Century where the oil was used for cooking.

In 2016, the global output of sunflower oil was almost 16 million tonnes.

The oil contains approximately 59% polyunsaturated fat, 30% monounsaturated fat & 11% saturated fat. The seeds contain 40% oil (by mass).

This is currently under trials for use as a neat SVO fuel as it gels at around -12°C (54°F), which is better than rapeseed oil which freezes at -10°C (14°F).

Approximately 275kg of oil can be produced per acre.

Figure 27 Helianthus annus (Public Domain © 1859)

Following the extraction of oil from the seeds, the waste is typically either used as animal feed, or pressed into pellets to be used as biomass fuel. As a bio fuel, it has very similar properties to soybean.

Jatropha curcas – jatropha

This is native to the American tropics, but is now widespread throughout the world in many subtropical locations. There are many species, some are toxic, many are not.

The seeds contain between 25 – 40% oil & there is typically an extraction level of 80%. This oil is made up of 20% saturated fat & 80% unsaturated fat. The plant has a potential of producing 1,600 litres of biodiesel per acre, per year.

Because of this, in 2007 Goldman Sachs cited the plant as one of the most promising for future biodiesel production.

The oil, after transesterification processing meets the current EEC requirement for diesel fuel.

Figure 28 Jatropha curcas (Public Domain © 1863)

It can produce four times as much fuel as per acre than soybean & ten times as much as corn, but the downside is that it takes ten times as much water to raise a healthy crop. Currently, the seeds are processed into fuel in Brazil, Pakistan & the Philippines. Many other developing countries have small scale projects.

Zea mays – corn

This was first domesticated in Southern Mexico approximately 10,000 years ago. Currently, throughout the world, by weight, more corn is produced than any other grain.

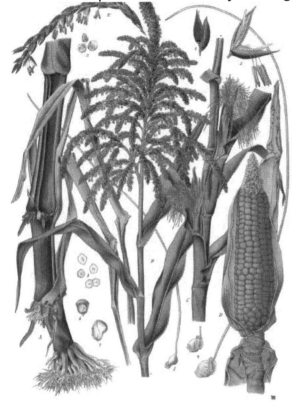

In 2014, 361 billion tonnes was produced globally.

Corn is processed into ethanol in many countries & is even mixed into petrol supplies in the UK & EEC with the aim of reducing pollutants.

Corn is cheap to produce; therefore 80% of the final cost is for transportation, production & packaging. Corn is also a foodstuff; therefore the cost will always be high, as demand is high.

Figure 29 Zea mays (Public Domain © 1897)

Corn is the world's third most important grain & can be grown in latitudes between sea level & 3,000 meters from South America to Southern Canada. It is also a successful crop throughout the world when planted in similar conditions.

In 2009, 85% of the corn planted in the USA was genetically modified. In addition, 40% of the corn now grown in the USA is processed into ethanol to be used as an additive for petrol, but the oil can just as easily be processed into biodiesel instead.

Animal derived FAME's

The cost of animal fat is far lower than that of vegetable oils & in the USA, 33% of all fat is from animal origin. The cost is lower as it is not classed as being edible for humans. Therefore it is used in pet food, soap making & in industry.

Animal fats are saturated. Beef & pork lard contain approximately 40% saturated fat, whilst chicken fat is around 33%. As a comparison, soybean oil is about 14% saturated & canola is around 6%. Beef & pork lard is therefore solid at room temperature. Whilst chicken or turkey fat tends to be a very thick liquid at this temperature. Therefore any biodiesel made from animal fats will only be suitable for use in hotter climates. It would not be suitable for use in Northern Canada or Northern Europe.

It was already stated in chapter 1 that in the EEC, the current minimum standard cetane number (CN) is 51 for petrochemical diesel, whilst premium petrochemical diesel fuel will be in the region of 60 CN. Soybean biodiesel is around 50 CN making it similar to normal petrochemical diesel, but animal fat biodiesel can be as high as 60 CN, therefore waste animal fat can produce excellent quality biodiesel that will help starting the engine & run much more quietly. It also produces slightly less amounts of nitrogen oxide (NOx) than biodiesel so it is typically in the range of that produced by petrochemical diesel. This is because an increase in the CN lowers nitrogen oxide (NOx) emissions by lowering the temperature in the early part of the combustion process.

Technically, any animal fat based fuel will have greener credentials than its plant made counterpart as it is a by-product, not a virgin material, therefore a saving of 50 – 80% on the carbon footprint can be achieved when compared to petrochemical diesel. Also, any fuel made from animal fat is generally known as 'renewable diesel', not 'biodiesel'. It has a different name as the process for converting it into useable fuel is different.

Biodiesel is created by using alcohol to react with the plant based oils with the aim of creating the biodiesel molecules. To create renewable diesel from animal fats, the fats are hydrogenated by reacting the fat with hydrogen at a high pressure & high temperature. This creates a pure, synthetic hydrocarbon molecule which is chemically identical to petrochemical diesel.

As the carbon footprint for the renewable diesel can be up to 80% lower than petrochemical diesel, some large companies are now experimenting with it commercially. The Finnish oil company Neste Oil now has four refineries making the fuel. Two in Finland, one in Singapore & one in the Netherlands. Many government & state fleets in the USA are also now starting to adopt renewable diesel purely because it can greatly help then reduce their carbon footprint with little or no effort.

There are however some problems. For instance, sulphur content can be a problem. In the USA, the sulphur content for fuel used on roads can not contain more than 15 parts per million, yet some beef & chicken based renewable diesel has been found to contain more than 100 ppm. Animal fat also contains phospholipids (a type of lipid that contains a phosphate group molecule in the overall molecule) which can cause insoluble particles to occur when they come into contact with water. These phospholipids can be a problem as they will deactivate the components in a diesel catalytic converter. Polymers can also be formed too during the high temperatures involved in the rendering. These polymers can increase the viscosity of the fuel which can in turn block pipes, jets & filters.

Therefore the process of creating renewable diesel will pose a challenge for any at home DIY biodiesel entrepreneur.

Algae derived FAME's

Algae based biofuels (or algal oil) are fuels created from algae, which are organisms that grow in aquatic conditions which use light & CO_2 to create biomass. Algae based oil is just like petrochemical diesel, it releases carbon dioxide (CO_2) into the atmosphere during combustion, but unlike petrochemical diesel, it releases only the carbon dioxide (CO_2) that was removed by photosynthesis during the algae's life. Petrochemical diesel releases carbon dioxide (CO_2) that was captured by plants & animals millions of years ago. As such, both algae based biofuels & plant based biofuels are both considered to be 'carbon neutral'. Petrochemical diesel (or any other fossil fuel) is not, because it will always add to the carbon already present in the atmosphere.

Currently, as there is a perceived food shortage in the world, some corporate & state bodies are researching whether algae can be grown commercially for fuel using land that is unsuitable for agriculture. The practice of algaculture does have some obvious benefits. It can be grown with a minimum impact on water sources as it can be grown using saline water & even wastewater. The algae is biodegradable & harmless to the environment if it is spilled. Algae does cost more per unit mass than other biofuel crops because of higher operating costs, but can yield between 10 – 100 times more fuel per unit area. Therefore if algae biofuel were to completely replace all the petrochemical fuel in the USA, then only one seventh of the area used in 2000 to grow corn would be needed.

At present there are government funded research projects in Australia, EEC, Middle East & New Zealand, to name just a few. There are also a number of corporations researching algaculture & November 2012, Solazyme & Propel Fuels made their first retail sales of their algae-derived fuel.

Algae can be converted into numerous types of fuels, depending on which technique & which part of the cell are used. The lipid, or oily part of the algae biomass can be extracted & converted into biodiesel through a process very similar to that used for converting any other vegetable oil. Some strains of algae can produce up to 60% of their mass as oil. This is due to the fact that they do not produce leaves, stems or roots, they just float free in water. This also speeds their growth. Overall they can out perform the oil palm by a factor of 23. Therefore, for each acre of oil palm that produces 100 litres, the algae will produce 2,300 litres. Apart from a water filled pond, only phosphorus (P) & nitrogen (N) are needed to act as food for the algae.

The algae naturally grows in ponds which act as photobioreactors. These ponds will operate best between 15 – 30°C (59 – 86°F) & are approximately 300mm deep. There does not need to be any expensive processing equipment involved, as the algae can be harvested by just scooping them off the surface of the water & then leaving them to dry in the sun. Once dried, they can then be processed in to biodiesel with the aid of hexane (C_6H_{14})(liquid). Another method if extraction is with an oil press (this method is called expression). By using the oil press & hexane, more than 95% of the available oil can be extracted. This method is therefore suitable for small scale projects in rural communities, but being scaled up to a large industrial complex has led to problems. To provide just 10% of the EEC's fuel needs would need ponds three times the size of Belgium. These would then need to be fed & the fertiliser needed would be equivalent to 50% of the EEC's current total land fertiliser. Therefore, any algae farming will only work on a small scale using local materials.

Harvests can be made from ponds every 12 weeks by adding a ferric-oxide power (iron powder) to the pond. This forms a ferric-oxide polymer that sinks to the bottom of the pond. After draining the pond, the biomass can be collected & dried.

There is a range of information online detailing all aspects of algae growing & production. A good place to start would be www.oilgae.com.

Conclusion

It can be seen that of the three sources of FAME's listed here, it will always be beneficial to use by-products rather than virgin materials to create any biofuel. Also, algae & plant sources will always be 'carbon neutral' when compared to fossil fuels.

Therefore the quest is to find a plant derived by-product. Luckily, there will be countless sources in your local town that will be ripe for harvesting.

Café's, restaurants & takeaways always use vegetable oil for cooking. They will also have to pay a company to remove their used oil which will be disposed of in landfill. This is therefore the perfect place to obtain used vegetable oil.

It will not be possible to choose what type of vegetable oil you want. It is a case of exploiting what is available. Café owners will thank you for taking their spent oil away as they have to pay for it to be removed legally & it just ends up finding its way into landfill. Therefore, by utilising this old, used vegetable oil, you are saving the café owner money, stopping the waste oil from going into landfill, then recycling the oil in a carbon neutral way, reducing the need for further petrochemical extraction from the ground.

This all sounds like a win, win scenario.

Chapter 5 – Environmental considerations

All the fuels that have been mentioned so far have a few things in common. Firstly, they are used to power engines. Secondly, there is always a by-product. That by-product is commonly referred to as pollution as it has an unwanted, negative effect on the environment. It is therefore worth looking into these effects & if at all possible attempt to reduce or eliminate them.

The composition of petrochemical diesel pollution

The diesel engine is claimed to be a heavy polluter. The phrase 'dirty diesel' is often thrown about, generally by people who think electric cars will save the world. It is therefore best to be informed before making any brash statements, so the best thing to do is look into subject & sort the facts from the fiction. So, to start, the average exhaust emissions from an average diesel vehicle are as follows[10]:

Nitrogen (N_2) – 75.2%
Oxygen (O_2) – 15%
Carbon dioxide (CO_2) – 7.1%
Water (H_2O) – 2.6%
Carbon monoxide (CO) – 0.043%
Nitrogen oxide (NOx) – 0.034%
Hydrocarbons (HC) – 0.005%
Aldehyde – 0.001%
Particulate matter (Sulphate & solid substances) – 0.008%

---

[10] Konrad Reif (ed): Dieselmotor-Management im Überblick. 2nd edition. Springer Fachmedien, Wiesbaden 2014, ISBN 978-3-658-06554-6. p. 171

The composition would be made up as follows:

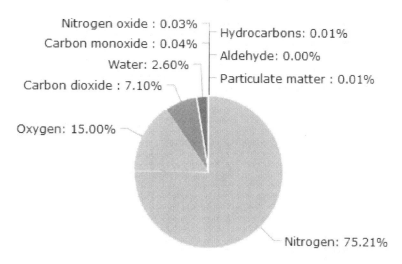

Figure 30 Diesel exhaust emissions (P Xavier © 2020)

The first is nitrogen (N). It is the seventh most abundant element in the universe. In atmospheric conditions two nitrogen atoms will bond forming dinitrogen (N_2) which is a colourless, odourless gas which makes up 78% of the Earth's atmosphere. That means it is the most common element on Earth. It also occurs in all organisms. The human body contains approximately 3% by mass. It is transparent to solar radiation & also infrared radiation, so is not a greenhouse gas. It is inert & it is non-toxic.

The next is oxygen (O). It is the third most abundant element in the universe. In atmospheric conditions two oxygen atoms will bond forming dioxygen (O_2) which is a colourless, odourless gas which makes up 20.95% of the Earth's atmosphere. The human body contains approximately 65% by mass (most of which is bound in water). It is transparent to solar radiation & infrared radiation, so is not a greenhouse gas. The human body is designed to breathe oxygen from the atmosphere at approximately 21%.

The next is carbon dioxide (CO_2). It is a gas that occurs naturally in the Earths atmosphere & is called a trace gas.

It makes up approximately 0.04% of the Earth's atmosphere. The level before the industrial revolution is estimated to be approximately 0.028%. It is a naturally occurring gas that comes from volcanoes, hot springs & when carbonate rocks are slowly dissolved by water. It is water soluble, so is found in groundwater, rivers, lakes, the ice caps, seawater & glaciers. It is also found in natural gas & petroleum. It is considered to be the primary source of carbon for all life on Earth, but is also considered to be the primary greenhouse gas. It is estimated that in 2018, carbon dioxide (CO_2) accounted for 81% of all greenhouse gasses[11], but just as humans & animals breathe oxygen, all plants breathe carbon dioxide. As all human & animals bodies store oxygen, all plants store carbon dioxide.

The next is water (H_2O). Water is an inorganic, tasteless, odourless & (almost) colourless chemical that is vital to all life on Earth. Water covers 71% of the Earth's surface, it is even in the air as a vapour. The vapour in the air is a greenhouse gas, but the lower the temperature, the less water vapour can be held in the atmosphere. All human, animal & plant life depend on water.

Carbon monoxide (CO) is the second most abundant element in the universe & is present in small amounts in the Earth's atmosphere where it is a colourless, odourless & tasteless flammable gas. It is toxic (>35 parts per million) to animals that use haemoglobin as an oxygen carrier in their blood. This therefore includes humans & the majority of mammals, but it is also naturally produced in human & animal bodies & believed to have a biological role. The largest source of carbon monoxide in the atmosphere is of natural origin & is made by photochemical reactions in the atmosphere, along with volcanoes & forest fires. Within one month of carbon monoxide (CO) being in the atmosphere, it is oxidised into carbon dioxide (CO_2).

---

[11] https://www.epa.gov/ghgemissions/overview-greenhouse-gases - 18/06/2020

Nitrogen oxide (NOx) is a generic term for the composition of nitrous oxide (N_2O) aka laughing gas, nitric oxide (NO) & nitrogen dioxide (NO_2).

These gases contribute to the formation of acid rain & smog. They are created when any fuel, including diesel fuel, is burnt at a temperature above 1,300°C (2,372°F).

Nitrous oxide (N_2O) is generally considered to be non-toxic & is used as an anaesthetic. It is even on the WHO's list of essential medicines. However, environmentalists claim it to be 298 times worse than carbon dioxide (CO_2), but nitrous oxide (N_2O) is not a greenhouse gas. It does contribute to the greenhouse effect because it destroys ozone in the upper atmosphere. It does occur naturally, such as during thunder storms, but it is estimated that 30% of the N_2O in the atmosphere is man made, & the majority of that comes from fertilisers that are used in agriculture. It is also used as a food additive (E942) & is generally found in aerosol cream where it is used as a propellant.

Nitric oxide (NO) is a colourless gas. It is a naturally occurring chemical in the human body & is also produced during the combustion in a diesel engine, when nitrogen combines with oxygen. In the atmosphere it can combine with water vapour to form acid rain & eats ozone in the upper atmosphere to create dioxygen (O_2).

Nitrogen dioxide (NO_2) is a reddish brown gas with a pungent, acrid odour when it is above 21.2°C (70°F). Below that temperature it forms into a yellowish brown liquid. Again this is formed naturally in nature & is produced by volcanoes & thunder storms. In increased concentrations, it has been shown to irritate airways in humans. In the atmosphere it can combine with water vapour to form acid rain & eats ozone in the upper atmosphere to create dioxygen (O_2).

Hydrocarbons (HC) are organic compounds consisting entirely of hydrogen & carbon molecules. This is basically un-burnt diesel fuel.

This happens because the molecule chain in diesel is long when compared to petrol. As such, it needs more oxygen to combust the fuel.

Aldehydes are organic compounds which are known irritants for the skin, eyes & respiratory tract. It is estimated that 20% of the aldehyde emissions is made up of formaldehyde, (CH_2O) which is a suspected carcinogen in humans.

Particulate matter (DPM) is also known as diesel exhaust particles (DEP). DPM includes diesel soot, aerosols, metallic particles, sulphates & silicates, all of which are known to be irritants for skin, eyes & the respiratory tract. As these particles are so small (most in the invisible sub-micrometer range of 100 nanometres) they are easily inhaled & then penetrate deep into the lungs where they can cause reflex coughing, wheezing & a shortness of breath.

The composition of straight vegetable oil (SVO) pollution

It is impossible to quantify the composition for straight vegetable oil as each batch can be derived from different feedstock, different farms, or even a hodgepodge of different plant oils. Therefore it is exactly the same question as how long is a piece of string?

However, in the previous chapter, ten of the most common plants used as feedstock for SVO were studied, therefore these will now be looked at regarding their potential pollution causing characteristics.

The oil bearing plants looked at earlier were: *arachis hypogaea* (peanut), *brassica napus* (rapeseed), *cannabis sativa* (hemp), *carthamus tinctorius* (safflower), *cocos nucifera* (coconut), *elaeis guineensis* (oil palm), *glycine max* (soybean), *helianthus annus* (sunflower), *jatropha curcas* (jatropha) & *zea mays* (corn).

It would be unlikely that all of these oils would be labelled in the supermarket. Some would be, but not all. Therefore unless you were to grow your own feedstock, you could not be 100% sure on the actual content of the oil. Even if the oil was labelled as corn oil (maize), it would be impossible to know from which farms, or area the oil was sourced from, or even which genus of *zea mays* the oil had came from, or even what the exact molecular makeup of the oil would be. It is likely therefore to be a mixture. As such, it would be impossible to give an exact figure for the level of emission gasses it could cause. Because of this, only rough estimates on the likely level of emission gasses are possible. This differs from petrochemical diesel as that is an engineered product, which is made to fine tolerances & exacting specifications.

Therefore, a great deal of data is produced because it is scientifically measured, controlled & monitored. However, as an example to how each of the ten example oils can differ, the following table should be examined.

CHAIN / OIL	12:0	14:0	16:0	16:1	18:0	18:1	18:2	18:3	20:0	20:1	22:0	22:1	24:0	24:1
Peanut	<0.1	<0.1	8.3 - 14	0 - 0.2	1.9 - 4.4	36.4 - 67	14 - 43	<0.1	1.1 - 1.7	0.7 - 1.7	2.1 - 4.4	0 - 0.3	1.1 - 2.2	<0.3
Rapeseed	<0.1	<0.2	2.5 - 7	<0.6	0.8 - 3.0	51 - 70	15 - 30	5 - 14	0.2 - 1.2	0.1 - 4.3	<0.6	<2	<0.3	<0.4
Hemp	<0.1	<0.1	7.7	0.2	2.4	13.3	56.7	13.6	0.9	0.8	<0.1	<0.1	<0.1	<0.1
Safflower	<0.1	<0.2	5.3 - 8.0	<0.2	1.9 - 2.9	8.4 - 21.3	67.8 - 83	<0.1	0.2 - 0.4	0.1 - 0.3	<1	<1.8	<0.2	<0.2
Coconut	45 - 53	17 - 21	7 - 10	3 - 6	2 - 4	5 - 10	1 - 3	<0.1	<0.1	<0.1	<0.1	<0.1	<0.1	<0.1
Oil palm	<0.5	0.5 - 2.0	6.5 - 10	<0.2	1 - 3	12 - 19	1 - 3.5	<0.2	<0.2	<0.2	<0.2	<0.1	<0.1	<0.1
Soybean	<0.2	<0.2	8 - 13.5	<0.2	2 - 5.4	17 - 30	48 - 59	4.5 - 11	0.1 - 0.6	<0.5	<0.7	<0.3	<0.5	<0.1
Sunflower	<0.1	<0.2	5 - 7.6	<0.3	2.7 - 6.5	14 - 39.4	48.3 - 74	<0.3	0.1 - 0.5	<0.3	0.3 - 1.5	<0.3	<0.5	<0.1
Jatropha	<0.1	0.1	14.2	0.7	7	44.7	32.8	0.2	0.2	<0.1	<0.1	<0.1	<0.1	<0.1
Corn	<0.3	<0.3	8.6 - 16.5	<0.5	<3.3	20 - 42.2	34 - 65.6	<2	0.3 - 1	0.2 - 0.6	<0.5	<0.3	<0.5	<0.1

Figure 31 Lengths of molecular chains expressed as percentages (P Xavier © 2020)

Figure 32 shows the molecular composition for each of the example feedstock oils. The oils are listed on the left & the figures along the top represent the length of the carbon chain (the first number) & then the number of double bonds (the second number), separated by the colon.

The figures in the table are all approximate percentages, therefore it can be seen that the molecules in the safflower oil can contain between 8.4 – 21.3% of molecules that are 18 molecules in length & contain one double bond.

It should also be clear from this table that the figures are all approximate & therefore any attempt to measure exhaust emissions when using any of these oils will result in a wide range of values which can change from minute to minute. The data is therefore at best only going to be approximate within a defined range.

The breakdown on the types of FAME's that are contained within these oils may therefore offer some clues as to the level of pollution that can be expected. Therefore the following chart shows how each of these oils are composed.

Oil \ Fats	Saturated	Monounsaturated	Polyunsaturated
Peanut	17	46	32
Rapeseed	7	63	18
Hemp	7	9	82
Safflower	7	14	79
Coconut	90	6	2
Oil palm	49	37	9
Soybean	16	23	58
Sunflower	10	20	66
Jatropha	20	48	6
Corn	13	24	59

Figure 32 Breakdown of oils expressed as percentages (P Xavier © 2020)

Again these figures are approximate due to the numerous factors that go in to the makeup of the oils. However, it is generally acknowledged that oils which contain the highest amount of unsaturated chains with one or two bonds will make the best oil for a fuel.

Therefore rapeseed (canola), hemp & safflower oils would be the preferred choice as a fuel oil, but it is known that the higher the level of polyunsaturated fat which an oil can contain means the higher the level of nitrogen oxide (NOx) will be produced in the exhaust gasses.

Therefore both hemp & safflower would be heavy polluters of nitrogen oxide (NOx) & that therefore just leaves rapeseed (canola) as the best option for a SVO fuel as it has a high amount of unsaturated fat & a low level of polyunsaturated fat. This then will have the added benefit of lowering the nitrogen oxide (NOx) emissions.

The composition of biodiesel pollution

Biodiesel is vegetable oil that has been chemically altered to be similar to petrochemical diesel. The vegetable oil then becomes thinner & therefore the viscosity becomes similar to petrochemical diesel. Vegetable oil is viscous as its molecules are attached to a glycerol molecule. The glycerol is therefore replaced & this liberates the FAME's which act as a replacement for petrochemical diesel. This makes perfect sense as the diesel engine is now tuned to diesel oil. Most of the individual components are diesel specific & therefore they will operate optimally when used with diesel.

Anything that has the same characteristics of petrochemical diesel will therefore perform much better in a diesel engine than something which does not.

It is possible to use SVO & convert it into biodiesel, but due to the costs (SVO is more expensive than petrochemical diesel), this is generally only ever undertaken by individuals as a way of practicing the conversion process. Most biodiesel which is added to the petrochemical diesel you buy at the forecourts (E7, E10 etc.) is made this way, but as cost is always an issue, most home biodiesel producers use solely waste vegetable oil (WVO) from cafés, restaurants & take-away's, then convert the WVO after thoroughly cleaning it.

Studies have found that nitrogen oxide (NOx) levels are slightly higher with biodiesel when compared to petrochemical diesel, although home biodiesel producers who make & study their own biodiesel have claimed that there is actually a reduction in emissions.

However, in this instance it should be assumed that there could be an increase of 10% in the level of nitrogen oxide (NOx) emissions.

Reducing pollution for petrochemical diesel, straight vegetable oil (SVO) & biodiesel

Over many years vehicle manufacturers have invested billions of pounds on developing solutions to reduce diesel emissions. Their investment has borne fruit as diesel engine produced carbon monoxide (CO) has fallen 82% from 1993 levels, diesel engine produced nitrogen oxide (NOx) has fallen 84% from 2001 levels & particulate matter (PM) has fallen 96% from 1993 levels.

These reductions have been achieved to meet the Euro standards for new cars. These standards began in 1992 & since their inception; each new standard has been more & more stringent. The vehicle manufacturers have always met the ever tighter emission controls as they have evolved over the years. Currently, all new diesel vehicles sold in the EEC are cleaner than they have ever been at any time in history. Diesel cars now emit, on average 15 – 20% less carbon dioxide (CO_2) than their petrol fuelled equivalents. This has led to the prevention of more than 3 million tonnes of carbon dioxide (CO_2) from going into the atmosphere over the past decade.

Euro standard	Introduction date		Emission limits		
	New approvals	New registrations	Petrol NOx	Deisel NOx	Deisel PM
Euro 1	01 07 1992	31 12 1992	0.97g/km	0.97g/km	0.14g/km
Euro 2	01 01 1996	01 01 1997	0.5g/km	0.9g/km DI	0.1g/km
Euro 3	01 01 2000	01 01 2001	0.15g/km	0.5g/km	0.05g/km
Euro 4	01 01 2005	01 01 2006	0.08g/km	0.25g/km	0.025g/km
Euro 5	01 09 2009	01 01 2011	0.06g/km	0.18g/km	0.005g/km
Euro 6	01 09 2014	01 09 2015	0.06g/km	0.08g/km	0.045g/km

Figure 33 Euro standard chart (P Xavier © 2020)

Currently, Euro 6 is being introduced progressively between September 2017 & January 2021, where all new cars are now being tested to meet the limits over a variety of real world conditions. This is to ensure that the latest level of emissions are reduced not only in the lab, but also on the road too. This was introduced as a result of the VW emission scandal where a 'cheat device' was fitted to ensure all their vehicles passed the emission tests.

Thanks to years of research, there are now multiple options available to help reduce diesel emissions.

DPF's are now so efficient that they can remove up to 99.9% of the particulate matter that is present in diesel exhaust emissions. Since 2011 when the Euro 5b exhaust emissions legislation was introduced, DPF's have now become pretty much mandatory for new diesel vehicles. A detailed explanation on how a DPF works was covered in chapter 1.

Methods of recirculating a portion of the exhaust gas back into the engine cylinders is also now employed. The first is known as 'exhaust gas recirculation' or EGR. This whole system will be controlled by the ECU with the aid of an EGR valve, which gives live feedback so the ECU can make whatever adjustments it deems necessary to reduce emissions.

The system redirects exhaust gasses over an intercooler to cool the gasses, then directs the gasses back into the cylinder. This can be up to 50% of the overall exhaust gasses, thereby greatly reducing the load on the pollution control system fitted in the exhaust pipe.

The downside to this system is that it can lead to excessive wear on the engine as the exhaust gasses will contain tiny carbon particles which can attach to the piston rings & also get into the oil, thereby increasing engine wear. The power output is also reduced as a result because it reduces the peak combustion temperature. This also introduces the drawback of burning less fuel in the power stroke, so, as a result, there is more unburnt fuel that will need to be recirculated into the cylinder. The whole EGR system makes the diesel engine less efficient.

Similarly, the reduction in temperature can also be achieved through the use of turbochargers as this also reduces the temperature in the cylinders with recirculated air. This reduction in temperature reduces the nitrogen oxide (NOx) emissions as it is the higher temperatures found in the diesel engine that creates these emissions. By recirculating a portion of the exhaust gas not only reduces these emissions by the reduction in temperature, but it also allows any unburnt diesel fuel in the exhaust gas another chance to combust in the cylinder on its second run through, which will also adds to the fuel economy.

Superchargers can also be used to reduce the temperature in the cylinders, but this method does not recirculate the exhaust gasses, therefore it is not as efficient as turbocharging.

Both turbocharging & supercharging will be aided in reducing the temperature on the air intake when an intercooler is included in the design. This is a small radiator type device which the gasses are passed through or over. The intercooler will then remove some of the heat & therefore aid the whole process of reducing emissions.

There are also what have become collectively known as NOx control technologies. The first of these is an SCR, where a catalyst reduces NOx to different chemicals; this is gaseous nitrogen (N_2) & water (H_2O) in the presence of ammonia (NH_3). This is achieved through the addition of AdBlue that has to be kept topped up.

Vehicle manufacturers use various names for this system. The particular model of the car may have SCR or blue in the name, but any diesel registered after September 2015 will need AdBlue. AdBlue is also manufactured under various names such as Bluedef, BlueTec etc., but they are all based on the same AdBlue/SCR technology.

The second is known as a 'lean NOx trap', 'NOx absorber' & abbreviated as a LNT. These work by allowing NOx to be adsorbed onto a catalyst (typically zeolite which is a microporus aluminosilicate mineral – aka cat litter) during lean engine operation (when there is a higher ratio of air to fuel). This NOx is then catalytically reduced during short periods of fuel-rich operation of the vehicle. This can be achieved by injecting diesel fuel onto the trap. The NOx in the trap then reacts with the hydrocarbons in the fuel to produce nitrogen (N_2) & water (H_2O). The VW Jetta TDI & VW Tiguan both use NOx traps & are marketed under the name BlueTec. These traps will be poisoned by sulphur oxide (SOx), which is adsorbed more readily into the zeolite than NOx. Therefore regular regeneration is needed, but this then greatly reduces the lifespan of the LNT.

There is also the DOC (as seen in chapter 1), which is the diesel engine specific catalytic converter. Overall, these systems & methods have helped reduce the emissions of new petrochemical diesel powered cars to extremely low levels. It would therefore be fair to say that they are now lower than they have ever been.

According to industry & governmental reports, the individual polluting chemicals contained in the exhaust emissions from SVO are lower than petrochemical diesel, except for the nitrogen oxide (NOx) emissions which can increase slightly. However, tests by individuals who use SVO as a fuel & also fuel co-operatives find that all emissions are reduced, including the nitrogen oxide (NOx) emissions.

If further savings are required for SVO emissions, exactly the same methods that are employed for petrochemical diesel will need to be employed. This will include the use of a 'diesel particulate filter' or DPF, 'exhaust gas recirculation' or EGR, turbocharging, supercharging, 'selection catalytic reduction' (SCR) with AdBlue (DOC) or another form of NOx trap. Also, as with any engine, keeping it serviced regularly & clean will also go some way in reducing the emissions.

Nitrogen oxide (NOx) is the main bone of contention with regards to biodiesel emissions as there is so much conflicting data & advice. But it is widely accepted within the fuel industry that a 10% increase in nitrogen oxide (NOx) emissions can be expected. There is however an easy solution to resolve any increases & that is to retard the engine timing between 1 – 3 degrees. This is achieved by matching the timing of the fuel injection to the combustion characteristics of the biodiesel. This will then reduce the nitrogen oxide (NOx) emissions without impacting the levels for any of other exhaust gasses. It will however slightly decrease the performance of the engine at the same time, but this is due to the process favouring engine emissions over engine performance. Any competent mechanic should be able to undertake this work on your behalf.

This method also has other drawbacks as well as reducing the engine performance. Because the engine was detuned from petrochemical diesel in favour of biodiesel, if petrochemical diesel was then used again as a fuel, the petrochemical diesel would be less efficiently burnt & therefore produce a higher level of emissions.

As a result, if 100% biodiesel (B100) is used at all times to fuel the engine, then retarding the timing is a great solution, but if only mixtures are used (for instance, B20, B50 etc.), or even just occasionally, then retarding the engine will make the emissions worse.

Therefore, the only solution would be to use a diesel CAT. Older cars that do not have a CAT fitted as standard may be able to retrofit a CAT. Fitting a CAT & reducing the engine timing will have the best overall result, but if a CAT is fitted & the engine is tuned to biodiesel, but petrochemical diesel is used as the fuel instead, the exhaust gasses will then contain unburnt fuel which will destroy the CAT.

When compared to an engine running petrochemical low-sulphur diesel, switching to biodiesel could see a 6% increase in nitrogen oxide (NOx) emissions. If the engine was also retarded, it would be reduced slightly to a 5% increase in nitrogen oxide (NOx) levels. If a diesel CAT were also fitted, then a reduction of approximately 30% could be expected.

Also, when biodiesel is compared to an engine running petrochemical low-sulphur diesel, there is a 15% reduction in carbon monoxide (CO). If a CAT is also fitted to the vehicle, a 97% reduction in carbon monoxide (CO) can be seen. But if the timing has been tuned to biodiesel, then this drops to a 94% reduction in carbon monoxide (CO).

Hydrocarbons (HC) can also be reduced when biodiesel is compared to an engine running petrochemical low-sulphur diesel. The biodiesel fuelled vehicle will see a 38% reduction. If a CAT were also fitted, then a 91% reduction would be the result, but if the timing was also tuned for biodiesel use, then this would reduce to a saving of 86% over the petrochemical low-sulphur diesel engine. A similar reduction will also result with the particulate matter (PM) when biodiesel is compared to an engine running petrochemical low-sulphur diesel. The biodiesel fuelled vehicle will see a reduction of 32% over the petrochemical engine. If a CAT were also fitted, then a 68% reduction would be the result, but if the timing was also tuned for biodiesel use, then this would reduce to a saving of 49% over the petrochemical low-sulphur diesel engine.

This is not all - volatile organic compounds (VOC's) emissions will reduce by approximately 49%, polyaromatic hydrocarbon emissions will reduce by approximately 70%.

Biodiesel does not contain any sulphur, therefore its emissions do not contain the acid rain causing sulphur dioxide (SO_2). Carbon dioxide (CO_2) emissions are cut by up to 100%, polycyclic aromatic hydrocarbon (PAH's) emissions are reduced by up to 97% & aldehyde emissions are reduced by up to 13%.

Therefore, overall, biodiesel outperforms other fuels in all areas except nitrogen oxide (NOx) emissions, but provided the engine is maintained & steps are taken, these nitrogen oxide (NOx) emissions can be reduced to acceptable levels.

Even more savings

When a lifecycle analysis is made on the impact of SVO fuel, it has found that it pollutes approximately 74% less than petrochemical diesel. This is due in part to the fact that all the carbon dioxide (CO_2) released from SVO was only held temporarily by the fuel. As it will only release what was stored by the plant during its life. Petrochemical diesel on the other hand will release the carbon that was absorbed millions of years ago.

The SVO will therefore neither add nor remove carbon dioxide (CO_2) into the atmosphere. The petrochemical diesel will just add carbon. SVO is therefore classed as being carbon neutral. SVO fuels will therefore always be carbon neutral; however, transporting oil from South America to Europe will tip the balance to petrochemical diesel being more carbon efficient, therefore it is important that any oil that is produced remains local & not sent off to a different continent to be used.

Biodiesel performs even better. As WVO is a byproduct which is recycled rather than it finding its way into landfill there are numerous benefits. In addition, the carbon dioxide (CO_2) released during combustion of biodiesel is carbon negative when compared to petrochemical diesel.

For every litre of biodiesel that is used instead of petrochemical diesel, 2.2kg of carbon dioxide (CO_2) is saved from going into the atmosphere.

In addition, biodiesel is biodegradable & non-toxic, therefore the likelihood of environmental disasters caused by oil spills is reduced to zero. Biodiesel is less toxic than table salt & as biodegradable as sugar. It also has a high flashpoint (150°C - 302°F), so the biodiesel will need to be at this temperature before it will ignite. Petrochemical diesel has a lower flashpoint (52°C - 126°F). Biodiesel is therefore safer to store & to transport. Biodiesel also lubricates the engine which reduces wear. Petrochemical diesel does not lubricate.

Biodiesel is an infinite resource. Vegetable oils can be grown & harvested every year. But there is only a limited supply of petrochemical fuels. It has been known for some time that the world has reached 'peak oil' (the maximum point after which there are only dwindling reserves left to tap underground). Therefore the supply of petrochemical fuels will steadily decrease, whilst the cost will steadily rise.

Petrochemical fuels also contribute towards many countries national debt as the majority of countries are net importers of petrochemical fuels, not net exporters.

The UK, despite producing 1.5 million barrels of crude oil per day in 2009, consumed 1.7 million barrels per day, making the UK a net importer of crude oil. The UK's North Sea oil fields reached 'peak oil' in 1999 & so afterwards production has just declined. As a consequence, the UK has been a net importer of crude oil since 2005.

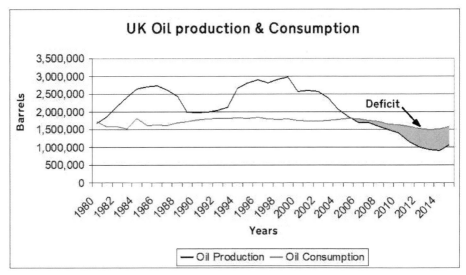

Figure 34 UK oil consumption (P Xavier © 2020)

Also, despite the UK still having the largest proven crude oil reserves (3.1 billion barrels) in the EEC, the trend for importing more oil than exporting it is set to continue & also increase, along with an increasing trade deficit. In 2013 crude oil imports accounted for 1% of the national debt. Since then, its percentage has risen by 3.8% every year.

It is not just the UK that is a net importer of crude oil, there are numerous other countries too. Also, even if your particular country is not a net importer, but a net exporter, by making biodiesel from WVO, the equivalent volume of petrochemical diesel is saved from the domestic market, therefore the economy benefits as the petrochemical diesel which was saved from the domestic market can now be sold overseas & that will result in greater profits for your country's exports & hopefully the less taxes the government will need to raise from you, the consumer.

Biodiesel is therefore good for the environment, the economy, the engine & your pocket. Rarely is there such a clear cut example of a winning strategy.

Chapter 6 – Modification or transformation

As has been seen earlier, modern diesel engines are now designed to run purely on crude oil derived diesel fuel. Therefore modifications must be made to the engine if any alternative fuels are to be used to fuel it. The simple choice is to either modify the engine to suit the fuel, or modify the fuel to suit the engine. This chapter therefore looks at modifying the engine & modifying alternative fuels.

Engine modification

Engine modification is not a complicated business. Anyone with mechanical skills should be able to undertake an engine modification. If DIY is not a strong point, then there are many specialist companies that will undertake the modification if your local mechanic does not wish to carry out the work. The benefit of modifying a diesel engine in a vehicle is that it will allow the engine to run on straight vegetable oil (SVO). This is clean unused vegetable oil such as the bottles & drums of cooking oil that you will see in the supermarket. However, the vehicle will need a dual fuel system. One diesel fuel tank & the associated pipe work to get the fuel to the engine & also an alternative fuel tank for the SVO, along with another set of pipe work to get the SVO fuel to the engine.

The whole system must be switchable (so that the driver can chose which fuel source to use) & it must always be switched to diesel before switching off the ignition, so that the engine will be purged of SVO & primed with diesel ready to restart the engine with diesel when the engine is cold. If this is not done, the SVO could solidify in the engine, blocking jets & pipe work.

This problem is caused by the viscosity of the SVO. As previously stated, SVO is much more viscous than diesel fuel. Therefore the viscosity will need to be reduced. This can be achieved temporarily by heating the SVO. This is because when the SVO has reached the desired temperature, it will have a similar viscosity to petrochemical diesel fuel, if it is too cold, it is far too viscous & will therefore block jets, filters & pipes. Also, heating the SVO to attain the correct viscosity will also allow the SVO to achieve good combustion. The temperature of the fuel being injected into the combustion chamber must be at least 71°C (160°F).

Heating the SVO can be achieved through a number of methods. The first of which is by using a heat exchanger, of which there are two types. The flat plate & the coaxial. Both types offer similar efficiencies, but the flat plate type is smaller & will therefore cost far more. This method is usually known as the 'final fuel heater', as it is the very last part of the heating system before the SVO is injected as the fuel. It is used in conjunction with the other heating methods.

The next method employed to heat the SVO is to heat the fuel line. A typical method is known as the 'hose in hose' method, where the SVO fuel hose is routed through the inside of the hot coolant pipe work. Alternatively, a longer length of flexible coolant pipe work would be coiled around the outside of the SVO pipe work & then secured in position.

Heating the SVO fuel tank is also another popular option. This can be heated electrically, or by routing the hot coolant pipe work through or under the tank. However, it is important that the tank is not heated to 71°C (160°F). Filters can also be employed to heat the SVO, but these tend to be electrically heated.

Using electrical methods to heat the SVO will always be a drain on the alternator. Therefore if electrical heating is adopted, then a slightly more powerful alternator would be needed to cater for this extra electrical load that has been imposed on the system.

It is therefore prudent to recover as much heat as possible from the hot coolant pipe work to preheat the SVO. It would be less of a strain on the alternator if only the last 10% of the heating were to be achieved electrically, rather than 100%. It is more efficient & will have less impact on the engine. Although there is a simple equation that can be used if you find that you need to select another alternator.

$$\text{Upgrade rate} = \text{Generic rate} \times 1.5 + \left(\frac{\text{Watt draw of new parts}}{14} \right)$$

Some of the more advanced (& more expensive) conversions include more heaters. It is possible to convert your engine without heaters & that would be permissible if you plan to drive solely in the tropics during the summer months, but if not, methods of heating the SVO must be employed. Therefore, for the majority of individuals, a two tank system would be required.

With any two tank system, the engine will have three fuel modes. Diesel only mode, SVO only mode & purge mode. As stated previously, the engine should always be started in diesel mode. It should not be switched to SVO mode until the engine has reached operating temperature & the SVO has been heated adequately to achieve the required 71°C (160°F). The switching can be achieved either manually or electronically & the current operating mode should be indicated to the driver/operator at all times, so that they will know when best to switch to purge mode. Automatic switching is the preferred option, so as to remove the chance of switching to SVO too early.

During diesel only mode the return fuel is sent to the diesel fuel tank. During SVO mode, the returning fuel is sent to the SVO fuel tank. During purge mode, fuel is drawn from the diesel tank & the returning fuel is sent to the SVO tank. These modes are possible due to a two way valve fitted on the fuel return line.

Standard fuel pumps found in diesel engines are designed to pump diesel. Therefore an internal gear pump is the best option as if the SVO is cold & therefore too viscous, the increased load will quickly break any diaphragm, vane or solenoid pump. Therefore, a separate fuel pump may be needed, especially if the standard diesel pump is located in the fuel tank, or in another inaccessible location.

Any materials used in the SVO conversion must also be considered as SVO will react to certain materials & will affect others. For instance, copper, steel, iron, mild steel or brass will breakdown the SVO & it may become unusable as fuel. It is estimated that up to 33% of all SVO conversions have suffered from this at some point. It is therefore important to check all washers & components which are contained in the fuel system. High grade stainless steel & aluminium components should be used as alternatives. Heat exchangers typically have copper components as copper is extremely good at heat transfer, therefore selecting a suitable alternative may prove problematic.

Also, SVO is very unstable (when compared to petrochemical diesel) as if it is exposed to air, steam, sunlight, certain enzymes or a high heat, the SVO will breakdown & become unusable. This will lead to the SVO clouding & then blocking the jets & filters. This is called oxidative polymerisation. It is the same chemical process that occurs with oil based paint. This is when the surface hardens. This is fine with paint, but not with SVO. With SVO, it increases the viscosity. It will also happen naturally over time with SVO, therefore it is always advisable to only have up to two weeks worth of SVO in the tank at any one time. The only way to ascertain if the oil has been affected by oxidative polymerisation is to look, smell it or taste it. If it's good enough to drink or cook with, then it is good to use as a fuel. If it smells rancid or is unpalatable, then it is not suitable to be used as fuel. Also, as the SVO oxidises, it will become darker in colour.

Unsaturated oils always contain a high amount of molecular double bonds, but it is the presence of these bonds that leads to SVO instability. Saturated oils are therefore more stable, but tend to be solid at room temperature, so a compromise is found with cooking oil. It has enough unsaturated oil to be a liquid, but not enough saturated oil to be solid at room temperature. There will also be a problem with the SVO deteriorating any rubber that it will come into contact with. This could therefore mean using washers & hoses made from a fluroelastomer material such as Viton.

Therefore if converting the engine seems to be a sure way of storing up future problems for the engine, it may be possible to change the engine for one that can use SVO as a fuel, without any modifications. The Peugeot Citroën XUD engine is capable of running unmodified on SVO, as was seen in chapter 1, but it is not the only engine with this ability.

Elsbett engine

Another engine that is powered by SVO & WVO is known as the Elsbett engine (aka Elko engine from **El**sbett **Ko**nstruktion). This engine was conceived by Ludwig Elsbett in 1977 & was designed to use SVO & WVO as a fuel source, which it did most successfully. It was 37% more efficient than the nearest petrochemical diesel powered engine at the time.

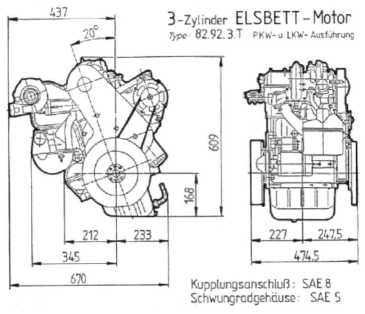

Figure 35 Elsbett engine (Elsbett 1992)

Unfortunately these engines were never manufactured on a large scale & no companies currently manufacture them, but over 1,000 of these engines were manufactured between 1979 – 1994, so there may still be some around that could appear for sale from time to time. For everyone else who can not buy a XUD or Elsbett engine, converting a diesel engine will be the only way forward for anyone who wishes to fuel their diesel with SVO or WVO.

But if changing the engine or converting it is not an option, then converting the SVO or WVO will be the only choice. It also happens to be the cheapest choice too.

Fuel modification

As was stated earlier, modification of the fuel will allow the use of biodiesel in an unmodified diesel engine. This is achieved with a chemical process called transesterification. This process was first undertaken in 1853 by J Patrick & E Duffy which was many years before Dr Diesel even invented his first diesel engine.

The application of transforming vegetable oils into a fuel is credited to G Chavanne who in 1937 was granted a patent in Belgium. His patent was entitled *'Procedure for the transformation of vegetable oils for their uses as fuels'*. That process which Chavanne had set out in 1937 has not altered & is now the accepted method for creating biodiesel.

There are numerous methods of transesterification, but the one that is of interest to the biodiesel maker is called base-catalysed transesterification.

Base-catalysed transesterification

All esters are created by a process called esterification, which is a chemical reaction in which two reactants form an ester. In the production of biodiesel, a type of esterification is employed called 'base-catalysed transesterification', where lipids (fats & oils) are reacted with an alcohol (typically methanol or ethanol) which in turn forms biodiesel & glycerol. This is possible because vegetable oils & animal fats are predominantly made from triglycerides.

The triglyceride molecule is made from three connected chains of fat. When this molecule interacts with methanol (CH_3OH) molecules, which acts as the catalyst, the chains of fat are liberated as fatty esters & the methanol is transformed into a colourless, odourless viscous liquid that is non-toxic & sweet tasting called glycerol ($C_3H_8O_3$).

Triglyceride molecule + Methanol

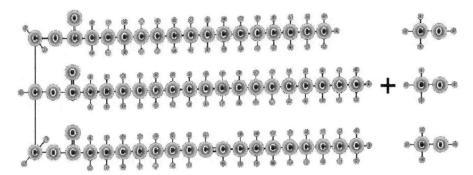

*Figure 36 Base-catalysed transesterification of triglyceride (P Xavier ©
2020)*

The fatty acid methyl esters are the molecules that make up
the biodiesel; the glycerol molecule is simply a byproduct of
the transesterification process. The triglyceride molecules
(cooking oil) have therefore now been transformed into
biodiesel (fatty acid methyl esters FAME's).

Fatty acid methyl esters   Glycerol

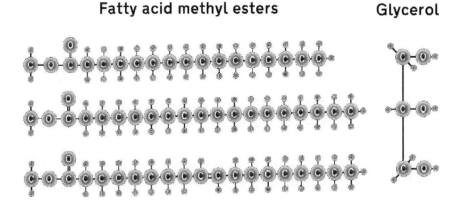

*Figure 37 Post base-catalysed transesterification of triglyceride (P
Xavier © 2020)*

As well as liberating the FAME's, the methanol molecules
have attracted some atoms from the triglyceride molecule &
have also become linked together, making another molecule;
in this instance glycerol. Therefore, adding methanol to the
triglyceride molecule has created the FAME's & glycerol.

The process is simply demonstrated in the following diagram.

Figure 38 Base-catalysed transesterification (P Xavier © 2020)

This process is completely reversible. In fact, as the methanol molecules are reacting with the triglyceride molecules, some of the resultant glycerol molecules & fatty acid methyl esters (FAME's) are reacting to reverse the process. The whole process will however reach equilibrium eventually. To minimise this reverse process happening, an excess amount of methanol is used, just enough to tip the balance enough to outweigh the reverse process.

When SVO is used for this process, it can be assumed that the majority of the oil is made from triglyceride molecules; therefore the process is exactly as has been demonstrated. However, there is a slight issue if WVO is used because when the cooking oil has been repeatedly heated, some of the triglyceride bonds will break, which results in a number of the fatty acid chains breaking free. These are therefore known as 'free fatty acids' (FFA's) & it is these FFA's that are problematic because the presence of any FFA's in the mix will turn the biodiesel acidic, which will then in turn damage the engine. Therefore these FFA's need to be transformed into soap.

This soap byproduct will then mix with the glycerol byproduct & the whole process can be accomplished during the transesterification process. This additional processing is called **titration** & typically either sodium hydroxide (NaOH) or potassium hydroxide (KOH) is used to accomplish this transformation when making biodiesel.

Chapter 7 – Danger, do not touch

DO NOT SKIP OVER THIS CHAPTER

One of the things that should have become obvious in the previous chapters is that there has been a lot of talk about chemicals & chemical reactions. Also, as the majority of individuals are not chemists or have had much experience with 'chemistry' since leaving school, it is important to look at some of these chemicals as some are dangerous & can therefore pose a real danger to people, property & the environment.

Protective equipment (PPE) should be used at all times, correct storing & handling of these chemicals should be also be observed at all times. It is also imperative that these chemicals are correctly labelled at all times & stored securely, away from children & animals. Measuring & mixing should also be undertaken with great care & in a controlled environment where any spills or splashes are dealt with immediately to ensure that the spillage is not walked into a house where it has the potential to cause serious injury to people, property & the environment.

It is also advisable that if the production of biodiesel is undertaken with any frequency, then a small laboratory should be constructed, along with a fireproof chemical store & an external concreted area with a bund wall constructed for the storage, mixing & processing of the chemicals & the oils.

If stored & handled correctly, there is less chance of accidents occurring, but it should be remembered, there will always be a degree of risk involved & therefore **serious injury & even death can be the result**.

DO NOT SKIP OVER THIS CHAPTER

Methanol

Methanol[12] (CH_3OH) aka methyl, methyl alcohol & wood alcohol is a very dangerous, volatile chemical. When making biodiesel, technical grade methanol is used which is 98 – 99% pure.

Methanol (CH_3OH)

DANGER

At room temperature it takes the form of a clear, light, volatile, colourless, highly flammable liquid that has a distinctive odour similar to drinking alcohol.

If ingested, just 10ml of methanol can cause permanent blindness & 30ml can cause death. On the human body it will act as a central nervous system depressant & also the liver will quickly convert methanol into formaldehyde. The liver will then turn the formaldehyde into formic acid which is toxic.

Methanol is also highly flammable & burns with an invisible flame, which means any potential spills & splashes could auto ignite & be unnoticed. Clothes too could catch alight if anyone were to bend over burning methanol. Methanol also kills nerve cells, so it would be possible that an individual may catch fire & not even know they were burning. The fumes from methanol are not only highly flammable, but can also be quickly absorbed through skin & cause intoxication. Therefore no one should ever bend over any vessel which contains methanol; it should only be used outdoors, with the use of fireproof gloves, a fireproof long sleeved top, goggles, a face shield & a fireproof apron. A typical face mask will offer zero protection from the methanol fumes, therefore a mask suitable for methanol use would be a suitable powered air-purifying respirator (PAPR) with a chemical cartridge. It is also advisable to use a hand pump to transfer methanol from one container to another, so as to minimise the risks that methanol presents.

---

[12] www.inchem.org/documents/icsc/icsc/eics0057.htm - 07/05/2020

As methanol vapour is heavier than air, it can collect in low lying pockets & remain a danger for some time. The vapour can even flow down a drain, travel a great distance & be a danger elsewhere. A source of running water should also be at hand anywhere that methanol is used. A hose should be used immediately to wash down any area that has come into contact with methanol. Methanol will dilute in water, but even at a level of 75% water & 25% methanol, it is still considered to be a flammable liquid, therefore copious amounts of water are needed to dilute the methanol down to a safe level.

Methanol should only be stored in clean containers made from either mild steel, stainless steel, high density polyethylene or vulcanized natural rubber. Methanol should never be stored indoors & should be stored in a bunded area (at least 110% bigger than what it is containing) to contain any spillages. Grounding is also advisable for the storage area, to reduce the risk of static electricity causing an explosion or methanol fire. Also, mobile phones, laptops, tablets, battery operated torches & any other battery operated device has the potential to cause an explosion, therefore should not be used in the vicinity of methanol.

Smoking should never, ever occur anywhere near methanol, or should the use of a naked flame. Therefore, electrical heaters should be used in the production of biodiesel, not gas flames.

If a methanol fire does occur, apart from water, use alcohol resistant aqueous film forming foam (AR-AFFF) which contains at least 6% foam.

Further information on methanol can be obtained from European chemical Substances Information System (ESIS) within the EEC. In the UK from the government website[13].

---

[13] www.gov.uk/government/publications/methanol-properties-incident-management -and-toxicology - 03/07/2020

In the USA from the government website 'Centres for Disease Control'[14] & in Canada from the safety council[15].

This is not an exhaustive list, a simple internet search will glean many other sources of information & safety sheets.

The 'globally harmonised system of classification & labelling of chemicals' (GHS), have issued the following hazard statements regarding methanol.

> H225 – Highly flammable liquid & vapour
> H301 – Toxic if swallowed
> H311 – Toxic in contact with skin
> H331 – Toxic if inhaled
> H370 – Causes damage to organs

They have also issued 37 precautionary statements regarding methanol.

If methanol is swallowed, 4ml would be lethal for a one year old child, 6ml would be lethal for a 3 year old & between 10 – 30ml would be lethal for an adult. Interestingly, astronomers have found that there is a huge cloud of methanol located approximately 6,500 light years from Earth in a region of space called W3(OH). This cloud of methanol is 288 billion miles across & can be found in the constellation of Cassiopeia.

First aid for methanol

If any individual has inhaled methanol vapour, the individual must be moved to fresh air if it is safe to do so & the individual must be kept warm.

---

[14] www.cdc.gov/niosh/ershdb/emergencyresponsecard_29750029.html - 03/07/20

[15] www.canadasafetycouncil.org/methanol/ - 03/07/2020

If the individual has difficulty breathing or if breathing has stopped, immediately start CPR & call for an ambulance.

If skin comes into contact with methanol, it should be flushed with copious amounts of fresh water for at least 15 minutes. If clothing or footwear has been in contact with methanol, then they should be removed under a shower. Wash any contaminated clothing or footwear before next use. If any irritation, pain or signs of toxicity occur, seek immediate medical attention.

If splashed into an eye, the eye should be washed with copious amounts of water for at least 15 minutes. The eyelid should be held open during this procedure to ensure all parts of the eye are washed. Following this, seek immediate medical attention.

If ingested, it is usually terminal. Symptoms can be delayed for 18 to 24 hours after ingestion. Do not induce vomiting, get medical attention immediately & insist the individual is hospitalised, monitored & treated for at least seven days.

Sodium hydroxide

Sodium hydroxide[16] (NaOH) aka lye, soda lye or caustic soda which is highly caustic & can cause severe chemical burns. It has an odourless, colourless crystalline structure (sold as granules) that will readily absorb water & this process liberates an extreme amount of heat which can also cause burns.

Sodium hydroxide (NaOH)

DANGER

Sodium hydroxide will absorb moisture from the air & individuals can become exposed by inhalation, ingestion & through skin or eye contact.

---

[16] www.inchem.org/documents/icsc/icsc/eics0360.htm - 05/07/2020

It is used as an ingredient in soaps & detergents, as well as drain cleaners. Exposure can cause skin & eye burns, skin & eye irritation, irritation to mucous membranes, pneumonitis & even a temporary loss of hair.

Sodium hydroxide is also corrosive to some metals, such as aluminium, tin, lead & zinc which produces the combustible & explosive gas hydrogen (H). It will also react violently with acids. If it comes in contact with ammonium salts it will produce ammonia.

Exposure to water or water vapour will produce heat. Never add sodium hydroxide to water. Add water slowly to the sodium hydroxide.

The European Chemicals Agency has issued advice on sodium hydroxide[17], as has the UK government[18]. In the USA information can be sought from the government website 'Centres for Disease Control'[19] & in Canada from the Canadian Centre for Occupational Health & Safety[20].

The 'globally harmonised system of classification & labelling of chemicals' (GHS), have issued the following hazard statements regarding sodium hydroxide.

H290 – May be corrosive to metals
H314 – Causes severe skin burns & eye damage

They have also issued 5 precautionary statements regarding sodium hydroxide.

---

[17]   www.echa.europa.eu/substance-information/-/substanceinfo/100.013.805   - 06/07/2020

[18]   www.gov.uk/government/publications/sodium-hydroxide-properties-uses-and-incedent-management - 06/07/2020

[19] www.cdc.gov/niosh/topics/sodium-hydroxide/default.html - 06/07/2020

[20]   www.ccohs.ca/oshanswers/chemicals/chem_profiles/sodium_hydroxide.html   - 06/07/2020

Interestingly, a popular food in some Scandinavian countries is called 'lutefisk'. It is made by soaking dried fish in sodium hydroxide until it turns to jelly. The jelly is then soaked in water for several days to remove the poison.

First aid for sodium hydroxide

If any individual has inhaled sodium hydroxide dust, the individual must be moved to fresh air if it is safe to do so & the individual must be kept warm. If the individual has difficulty breathing or if breathing has stopped, immediately start CPR & call for an ambulance.

If skin comes into contact with sodium hydroxide, any excess chemical should be brushed off & the area should be flushed with copious amounts of fresh water continuously for at least 60 minutes. If clothing or footwear has been in contact with sodium hydroxide, then they should be removed under a shower. Wash any contaminated clothing or footwear before next use. Seek immediate medical attention & if possible continue to wash the affected area whilst travelling to the hospital.

If splashed into an eye, any excess chemical should be brushed off the area, & then the eye should be washed with copious amounts of flowing water for at least 60 minutes. The eyelid should be held open during this procedure to ensure all parts of the eye are washed. Following this, immediately take to hospital for medical attention.

If ingested, rinse mouth with water, do not induce vomiting, but if vomiting occurs, have the individual lean forward to reduce the risk of aspiration. Rinse mouth with water again, then continually. Take to hospital to get immediate medical attention & insist the individual is hospitalised, monitored & treated, although if ingested, it is usually fatal.

Potassium hydroxide

Potassium hydroxide[21] (KOH) aka caustic potash, lye, potash lye, potassia & potassium hydrate. It is usually sold as either translucent pellets, flakes or a powder.

Potassium hydroxide (KOH)

DANGER

Potassium hydroxide will readily absorb water & this process liberates an extreme amount of heat which can also cause burns. It will even absorb moisture from the air.

It is almost the same as sodium hydroxide, but potassium hydroxide is more aggressive. It is used as an ingredient in soaps & detergents, as well as drain cleaners. Exposure can cause skin & eye burns, skin & eye irritation.

Potassium hydroxide is also corrosive to some metals, such as aluminium, tin, lead & zinc which produces the combustible & explosive gas hydrogen (H). It will also react violently with acids. If it comes in contact with ammonium salts it will produce ammonia. Vinegar however will neutralise potassium hydroxide, so it is worth keeping this to hand in a squeezable bottle to neutralise any spills or splashes.

Exposure to water or water vapour will produce heat. Never add potassium hydroxide to water. Add water slowly to the potassium hydroxide.

The European Chemicals Agency has issued advice on potassium hydroxide[22], as has the UK government[23].

---

[21] www.inchem.org/documents/icsc/icsc/eics0357.htm - 05/07/2020

[22] www.echa.europa.eu/substance-information/-/substanceinfo/100.013.802 - 06/07/2020

[23] www.gov.uk/government/publications/sodium-hydroxide-properties-uses-and-incident-management - 06/07/2020

In the USA information can be sought from the government website 'Centres for Disease Control'[24] & in Canada from the Canadian Centre for Occupational Health & Safety.

The 'globally harmonised system of classification & labelling of chemicals' (GHS), have issued the following hazard statements regarding potassium hydroxide.

> H302 – Harmful if swallowed
> H314 – Causes severe skin burns & eye damage

They have also issued 5 precautionary statements regarding potassium hydroxide.

Interestingly, potassium hydroxide is now commonly found in all washing powders as it is absorbed by the textile fibres during washing & then converts UV light (invisible to humans) into visible light which gives the textile fibre the appearance of reflecting more visible light, causing the clothes to look brighter & whiter.

First aid for potassium hydroxide

If any individual has inhaled potassium hydroxide dust, the individual must be moved to fresh air if it is safe to do so & the individual must be kept warm. If the individual has difficulty breathing or if breathing has stopped, immediately start CPR & call for an ambulance.

If skin comes into contact with potassium hydroxide, any excess chemical should be brushed off & the area should be flushed with copious amounts of fresh water for at least 15 minutes. If clothing or footwear has been in contact with potassium hydroxide, then they should be removed under a shower.

---

[24] www.cdc.gov/niosh/pel88/1310-58.html - 06/07/2020

Wash any contaminated clothing or footwear before next use. Seek immediate medical attention & if possible continue to wash affected area on way to hospital.

If splashed into an eye, any excess chemical should be brushed off the area, & then the eye should be washed with copious amounts of flowing water for at least 15 minutes. The eyelid should be held open during this procedure to ensure all parts of the eye are washed. Following this, immediately take to hospital for medical attention.

If ingested, rinse mouth with water, do not induce vomiting, but if vomiting occurs, have the individual lean forward to reduce the risk of aspiration. Rinse mouth with water again, then give small glass of water to drink. Take to hospital to get immediate medical attention & insist the individual is hospitalised, monitored & treated, although if ingested, it is usually fatal.

Sulphuric acid

Sulphuric acid[25] (H_2SO_4) aka oil of vitriol or hydrogen sulphate. It is viscous clear liquid, looking like oil, hence the old name of 'oil of vitriol'.

Sulphuric acid (H_2SO_4)

DANGER

Sulphuric acid will react with many metals, such as iron, aluminium, zinc, manganese, magnesium & nickel. The reaction produces the combustible & explosive gas hydrogen (H).

It is soluble in water when heated & used to produce fertilizers, drain cleaner & to clean metals. Exposure can cause skin & eye burns, skin & eye irritation. Breathing the fumes will also have serious consequences to health.

---

[25] www.inchem.org/documents/icsc/icsc/eics0362.htm - 05/07/2020

The European Chemicals Agency has issued advice on sulphuric acid[26], as has the UK government[27]. In the USA information can be sought from the government website 'Centres for Disease Control'[28] & in Canada from the Canadian Centre for Occupational Health & Safety[29].

The 'globally harmonised system of classification & labelling of chemicals' (GHS), have issued the following hazard statement regarding sulphuric acid.

H314 – Causes severe skin burns & eye damage

They have also issued 19 precautionary statements regarding sulphuric acid.

Interestingly, on the planet Venus, the upper atmosphere is made from sulphuric acid due to a photo-chemical reaction between carbon dioxide, sulphur dioxide & water vapour that is caused by the sun.

First aid for sulphuric acid

If any individual has inhaled heated sulphuric acid vapours, the individual must be moved to fresh air if it is safe to do so & the individual must be kept warm. If the individual has difficulty breathing or if breathing has stopped, immediately start CPR & call for an ambulance. Heated sulphuric acid vapours are VERY TOXIC & inhalation is usually fatal.

---

[26]   www.echa.europa.eu/substance-information/-/substanceinfo/100.028.768   - 06/07/2020

[27]   www.gov.uk/government/publications/sulphuric-acid-properties-incident-management-and-toxicology - 06/07/2020

[28] www.cdc.gov/niosh/topics/sulfuric-acid/default.html - 06/07/2020

[29]   www.ccohs.ca/oshanswers/chemicals/chem_profiles/sulfuric_acid.html   - 06/07/2020

If skin comes into contact with sulphuric acid, any excess chemical should be brushed off & the area should be flushed with copious amounts of fresh water for at least 30 minutes. If clothing or footwear has been in contact with sulphuric acid, then they should be removed under a shower. Wash any contaminated clothing or footwear before next use. Seek immediate medical attention & if possible continue to wash affected area on way to hospital.

If splashed into an eye, any excess chemical should be brushed off the area, & then the eye should be washed with copious amounts of flowing water for at least 30 minutes. The eyelid should be held open during this procedure to ensure all parts of the eye are washed. Following this, immediately take to hospital for medical attention.

If ingested, rinse mouth with water, do not induce vomiting, but if vomiting occurs, have the individual lean forward to reduce the risk of aspiration. Rinse mouth with water again. Take to hospital to get immediate medical attention & insist the individual is hospitalised, monitored & treated, although if ingested, it is usually fatal.

Glycerol

Glycerol[30] ($C_3H_8O_3$) aka glycerine, glycerin, propanetriol, 1,2,3-trihydroxypropane & also 1,2,3-propanetriol. It is a colourless odourless viscous liquid that is sweet to the taste & non-toxic.

This is a safe byproduct made during biodiesel production.

Interestingly glycerol is added to ice cream & chewing tobacco to improve texture & add sweetness without adding any sugar.

---

[30] https://echa.europa.eu/substance-information/-/substanceinfo/100.000.263 - 07/07/2020

It is also used as a base in toothpaste manufacturing to help maintain the shine & smoothness of the toothpaste. It is also burnt to power the diesel generators that provide the electricity for FIA Formula E race cars.

First aid for glycerol

There are no first aid points to note regarding glycerol as it is non-toxic.

Isopropyl alcohol

Isopropyl alcohol (C_3H_8O) aka 2-propanol, *sec*-propyl alcohol, IPA & isopropanol. It is colourless, highly flammable & has a strong odour that smells of a mixture of ethanol & acetone.

Isopropyl alcohol (C_3H_8O)

DANGER

Isopropyl alcohol is extremely flammable & is used in industry to make a variety of different products from air fresheners to hand sanitizers.

Isopropyl alcohol should always be stored in a flammable safety cabinet as its vapour will easily form an explosive atmosphere in the air. It should therefore also be kept in a sealed container, kept away from heat, sparks, flames or any other source of ignition at all times. It is also important to keep it away from strong oxidisers such as acetaldehyde, chlorine, ethylene oxide, acids & isocyanates. The next biggest danger that isopropyl alcohol posses is poisoning as it can enter the body by ingestion, inhalation & also absorption.

On the human body it will act as a central nervous system depressant & also the liver will remove 20 – 50% isopropyl alcohol from your body, but what remains will be broken down into acetone (C_3H_6O) which is a solvent that is a carcinogen & is also mutagenic.

It was on the United States Environmental Protection Agency's (EPA) list of toxic chemicals until 1995.

The fumes from isopropyl alcohol are not only highly flammable, but can also be quickly absorbed through skin & cause intoxication. Therefore no one should ever bend over any vessel which contains isopropyl alcohol; it should only be used outdoors, with the use of fireproof gloves, a fireproof long sleeved top, goggles, a face shield & a fireproof apron. A typical face mask will offer zero protection from the isopropyl alcohol fumes, therefore a mask suitable for isopropyl alcohol use would be a suitable powered air-purifying respirator (PAPR) with a chemical cartridge. It is also advisable to use a hand pump to transfer isopropyl alcohol from one container to another, so as to minimise the risks that isopropyl alcohol presents.

As isopropyl alcohol vapour is heavier than air, it can collect in low lying pockets & remain a danger for some time. The vapour can even flow down a drain & be a danger elsewhere.

A source of running water should also be at hand anywhere that isopropyl alcohol is used. A hose should be used immediately to wash down any area that has come into contact with isopropyl alcohol.

Isopropyl alcohol should only be stored in clean containers made from either mild steel, stainless steel, high density polyethylene or vulcanized natural rubber. Isopropyl alcohol should never be stored indoors & should be stored in a bunded area (at least 110% bigger than what it is containing) to contain any spillages. Grounding is also advisable for the storage area, to reduce the risk of static electricity causing an explosion or isopropyl alcohol fire. Also, mobile phones, laptops, tablets, battery operated torches & any other battery operated device has the potential to cause an explosion, therefore should not be used in the vicinity of isopropyl alcohol.

Smoking should never, ever occur anywhere near isopropyl alcohol, or should the use of a naked flame. Therefore, electrical heaters should be used in the production of biodiesel, not gas flames.

If an isopropyl alcohol fire does occur, apart from water, use alcohol resistant aqueous film forming foam (AR-AFFF) which contains at least 6% foam.

Further information on isopropyl alcohol can be obtained from European chemical Substances Information System (ESIS) within the EEC[31]. In the UK from the government website[32].

In the USA from the government website 'Centres for Disease Control'[33] & in Canada from the Centre for Occupational Health and Safety[34].

This is not an exhaustive list; a simple internet search will glean many other sources of information & safety sheets.

The 'globally harmonised system of classification & labelling of chemicals' (GHS), have issued the following hazard statements regarding isopropyl alcohol.

H225 – Highly flammable liquid & vapour
H319 – Causes serious eye irritation
H336 – May cause drowsiness or dizziness

They have also issued 5 precautionary statements regarding isopropyl alcohol.

---

[31]   https://echa.europa.eu/substance-information/-/substanceinfo/100.000.601   -   20/07/2020

[32] https://www.gov.uk/government/publications/isopropanol-incident-management - 20/07/2020

[33] https://www.cdc.gov/niosh/npg/npgd0359.html - 20/07/2020

[34] https://www.ccohs.ca/oshanswers/chemicals/flammable/flam.html - 20/07/2020

Interestingly, hand sanitizer can be made quickly by adding two parts isopropyl alcohol to one part aloe vera lotion, which is then mixed together thoroughly.

First aid for isopropyl alcohol

If any individual has inhaled isopropyl alcohol vapour, the individual must be moved to fresh air if it is safe to do so & the individual must be kept warm. If the individual has difficulty breathing or if breathing has stopped, immediately start CPR & call for an ambulance.

If skin comes into contact with isopropyl alcohol, it should be flushed with copious amounts of fresh water for at least 15 minutes. If clothing or footwear has been in contact with isopropyl alcohol, then they should be removed under a shower. Wash any contaminated clothing or footwear before next use. If any irritation, pain or signs of toxicity occurs, seek immediate medical attention.

If splashed into an eye, the eye should be washed with copious amounts of water for at least 15 minutes. The eyelid should be held open during this procedure to ensure all parts of the eye are washed. Following this, seek immediate medical attention.

If ingested, symptoms can be delayed for 18 to 24 hours after ingestion. Do not induce vomiting, get medical attention immediately & insist the individual is hospitalised, monitored & treated for at least seven days.

Phenolphthalein

Phenolphthalein ($C_{20}H_{14}O_4$) is a white or yellowish crystalline which is a weak acid & used as an indicator in chemistry & also as a laxative in medicine.

Phenolphthalein ($C_{20}H_{14}O_4$)

DANGER

Phenolphthalein was first made in Germany in 1871 by Adolf von Bayer & is colourless below pH8.0 & becomes pink at pH9.3. It then becomes a deep red above ph 9.6.

Phenolphthalein is decolorized by acids & is used extensively to check on the carbonation of concrete. When it is painted onto concrete, areas stained pink are alkaline & are therefore uncarbonated, whilst carbonated areas of the concrete will remain unstained, which indicates the problem areas. In the UK, the Concrete Society state that phenolphthalein is a substance of high concern & go on to recommend that it should only be handled with great care using safety gloves in a fume cabinet fitted with an extractor.

In the 1900's, phenolphthalein was used by the Hungarian government to colour wine as the native grape harvest failed that particular year, so they imported wine. The Hungarian people prefer deep red wines, so they coloured it with phenolphthalein to make it a deep red colour. The result was mass diarrhoea throughout the country. Shortly after in New York, a Hungarian pharmacist called Max Kiss heard about the wine's effect & decided that it may be a solution to constipation, so he added it to chocolate.

That chocolate became known as Ex-Lax which became the worlds best selling laxative of all time until it was banned in 1997 due to concerns that phenolphthalein was a carcinogen.

First aid for phenolphthalein

If any individual has inhaled phenolphthalein dust, the individual must be moved to fresh air if it is safe to do so & the individual must be kept warm. If the patient goes on to develop a cough, call for an ambulance. If the individual has difficulty breathing or if breathing has stopped, immediately start CPR & call for an ambulance.

If skin comes into contact with phenolphthalein, it should be flushed with copious amounts of fresh water for at least 15 minutes. If clothing or footwear has been in contact with phenolphthalein, then they should be removed under a shower. Wash any contaminated clothing or footwear before next use. If any irritation, pain or signs of toxicity occurs, seek immediate medical attention.

If splashed into an eye, the eye should be washed with copious amounts of water for at least 15 minutes. The eyelid should be held open during this procedure to ensure all parts of the eye are washed. Following this, seek immediate medical attention.

If ingested, do not induce vomiting. If person is conscious & alert, give 2 – 4 cups of water or milk. Get medical attention immediately.

Bromophenol blue

Bromophenol blue ($C_{19}H_{10}Br_4O_5S$) aka tetrabromophenolsulfonphthalein is an odourless pink or red powder that is used as a blue dye.

Bromophenol blue ($C_{19}H_{10}Br_4O_5S$)

DANGER

Bromophenol blue is used as an acid base indicator. It is yellow at pH 6 & blue at pH >7. It is effective in the range of pH 6 – 7.6.

The 'globally harmonised system of classification & labelling of chemicals' (GHS), have issued the following hazard statements regarding bromophenol blue.

H226 – Flammable liquid & vapour
H319 – Causes serious eye irritation
H335 – May cause respiratory irritation
H336 – May cause drowsiness or dizziness
H340 – May cause genetic defects
H360 – May damage fertility or the unborn child
H372 – Causes damage to organs through prolonged or repeated exposure
H373 –May cause damage to organs through prolonged or repeated exposure

They have also issued 30 precautionary statements regarding bromophenol blue.

First aid for bromophenol blue

If any individual has inhaled bromophenol blue dust, fumes, gas, mist, vapour or spray, the individual must be moved to fresh air if it is safe to do so & the individual must be kept warm. If the individual has difficulty breathing or if breathing has stopped, immediately start CPR & call for an ambulance.

If skin comes into contact with bromophenol blue, it should be flushed with copious amounts of fresh water for at least 15 minutes. If clothing or footwear has been in contact with bromophenol blue, then they should be removed under a shower. Wash any contaminated clothing or footwear before next use. If any irritation, pain or signs of toxicity occurs, seek immediate medical attention.

If splashed into an eye, the eye should be washed with copious amounts of water for at least 15 minutes. The eyelid should be held open during this procedure to ensure all parts of the eye are washed. Following this, seek immediate medical attention.

If ingested, do not induce vomiting. If person is conscious & alert, give 2 – 4 cups of water or milk. Get medical attention immediately.

Hydrochloric acid

Hydrochloric acid[35] (HCl) aka muriatic acid, spirits of salt, hydronium chloride or chlorhydric acid is the toxic water-based solution of hydrogen chloride gas that is highly corrosive when concentrated. As a liquid, it takes the form of a colourless (but almost slightly yellow) liquid with a pungent smell.

Hydrochloric acid (HCl)

DANGER

It is a strong inorganic acid that is mainly used in industry. The steel industry & the construction industry are the biggest users of this chemical. It is also used in a number of household cleaners.

Hydrochloric acid should be stored in a cool, dry, well ventilated area away from sources of moisture. It will corrode metals, therefore should be stored in glass bottles in a wooden cabinet. It should also be kept away from oxidizing agents, organic material, metals & alkalis.

Hydrochloric acid is corrosive to eyes, skin & mucous membranes. Breathing the fumes or mist will cause acute eye, nose & respiratory tract irritation & inflammation in humans. If the acid or mist comes into contact with skin, eyes or internal organs (such as lungs), the damage is usually irreversible & can be fatal in some cases. The United States Environmental Protection Agency (EPA) states that long term occupational exposure to hydrochloric acid has been reported to cause gastritis, chronic bronchitis, dermatitis & photosensitization, whilst prolonged exposure to low concentrations could cause dental decolourisation & erosion.

---

[35]   https://echa.europa.eu/substance-information/-/substanceinfo/100.210.665   - 29/07/2020

Since 1988, hydrochloric acid has been included in the *United Nations Convention Against Illicit Traffic in Narcotic Drugs & Psychotropic Substances* as it is used in the production of heroin, cocaine & methamphetamine.

Hydrochloric acid is available in concentrations of up to 38%. Industrial grade is 30 – 35%. It is often used in water to control its pH level.

A source of running water should also be at hand anywhere that hydrochloric acid is used. A hose should be used immediately to wash down any area that has come into contact with hydrochloric acid.

Further information on hydrochloric acid can be obtained from European chemical Substances Information System (ESIS) within the EEC[36]. In the UK from the government website[37].

In the USA from the government website 'Centres for Disease Control'[38] & in Canada from the Centre for Occupational Health and Safety[39]

The 'globally harmonised system of classification & labelling of chemicals' (GHS), have issued the following hazard statements regarding hydrochloric acid.

> H290 – May be corrosive to metals
> H314 – Causes severe skin burns & eye damage
> H335 – May cause respiratory irritation

---

[36]   https://echa.europa.eu/substance-information/-/substanceinfo/100.028.723   - 30/07/2020

[37]   https://www.gov.uk/government/publications/hydrogen-chloride-properties-incident-management-and-toxicology - 30/07/2020

[38] https://www.cdc.gov/niosh/topics/hydrogen-chloride/default.html - 30/07/2020

[39]

https://www.ccohs.ca/oshanswers/chemicals/corrosive/corrosiv.html?=undefined&wbdisable=true – 30/07/2020

They have also issued 8 precautionary statements regarding hydrochloric acid.

Exposure to 0.1% by volume to this gas in the atmosphere will cause death within just a few minutes.

Interestingly, it is hydrochloric acid that is in everyone's digestive juices within their stomach. If the body makes too much, it will cause gastric ulcers.

First aid for hydrochloric acid

If any individual has inhaled hydrochloric acid vapour, the individual must be moved to fresh air if it is safe to do so & the individual must be kept warm. If the individual has difficulty breathing or if breathing has stopped, immediately start CPR & call for an ambulance. Exposure for just a few minutes may result in a life threatening accumulation of fluid in the lungs (pulmonary edema) which will lead to death.

If skin comes into contact with hydrochloric acid, it should be flushed with copious amounts of fresh water for at least 15 minutes. If clothing or footwear has been in contact with hydrochloric acid, then they should be removed under a shower. Wash any contaminated clothing or footwear before next use. If any irritation, pain or signs of toxicity occurs, seek immediate medical attention. Even mild burns can cause permanent scaring.

If splashed into an eye, or even being exposed to the vapour could result in blindness, therefore the eye should be washed with copious amounts of water for at least 15 minutes. The eyelid should be held open during this procedure to ensure all parts of the eye are washed. Following this, seek immediate medical attention.

If ingested, it will cause severe burns to mouth & throat, oesophagus & stomach. This will cause difficulty with swallowing, intense thirst, nausea, vomiting, diarrhoea & death. Even a tiny amount ingested can cause death. Do not induce vomiting, get medical attention immediately.

Acetic acid

Acetic acid[40] ($C_2H_4O_2$) aka ethanoic acid, hydrogen acetate & methanecarboxylic acid takes the form of a colourless liquid. It smells like vinegar as (apart from water) it is the main component of vinegar (between 4 – 8% diluted in water).

Acetic acid ($C_2H_4O_2$)

DANGER

Acetic acid has many uses, including use as a chemical reagent, fungicide, herbicide, microbiocide, pH adjuster & as a solvent. It is also an excellent disinfectant as it kills mold & mildew.

Due to the disinfectant properties, it forms the basis for many 'green' environmentally friendly cleaning products along with most window cleaners & surface cleaners. Despite being present in all living organisms, breathing acetic acid can cause respiratory symptoms such as coughing, difficulty breathing, sore throat & nervous system problems such as dizziness & headaches. Contact with the eyes can result in burns, loss of vision, pain & redness. Contact with the skin can cause pain, redness & even burns. Ingestion can cause a sore throat, burning sensation, abdominal pains, vomiting, shock & collapse.

Acetic acid should be stored away from heat or sources of ignition. It should also be stored away from any oxidizing agents, reducing agents, metals, acids & alkalis.

[40]   https://echa.europa.eu/substance-information/-/substanceinfo/100.000.528   -
29/07/2020

It should always be kept in a cool, well ventilated place in a sealed container. In its pure form, ingesting just 210g will be enough to give a 50/50 chance of death.

Further information on acetic acid can be obtained from European chemical Substances Information System (ESIS) within the EEC[41]. In the UK from the government website[42].

In the USA from the government website 'Centres for Disease Control'[43] & in Canada from the Centre for Occupational Health and Safety[44].

This is not an exhaustive list; a simple internet search will glean many other sources of information & safety sheets.

The 'globally harmonised system of classification & labelling of chemicals' (GHS), have issued the following hazard statements regarding acetic acid.

> H226 – Flammable liquid & vapour
> H314 – Causes severe skin burns & eye damage

They have also issued 5 precautionary statements regarding isopropyl alcohol.

Interestingly, one of the most effective ways to wash fruit & vegetables is to use vinegar. Studies have shown that vinegar will kill 98% of the bacteria on fruit skins, whilst only an 85% bacterial kill rate can be achieved by brushing with soap & water.

---

[41]   https://echa.europa.eu/substance-information/-/substanceinfo/100.000.528   - 29/07/2020

[42]   https://www.gov.uk/government/publications/acetic-acid-properties-uses-and-incident-management/acetic-acid-general-information - 30/07/2020

[43] https://www.cdc.gov/NIOSH/NPG/npgd0002.html - 30/07/2020

[44]   https://www.canada.ca/en/health-canada/services/chemical-substances/fact-sheets/chemicals-glance/acetic-acid.html - 30/07/2020

First aid for acetic acid

If any individual has inhaled acetic acid vapour, the individual must be moved to fresh air if it is safe to do so & the individual must be kept warm. If the individual has difficulty breathing or if breathing has stopped, immediately start CPR & call for an ambulance.

If skin comes into contact with acetic acid, it should be flushed with copious amounts of fresh water for at least 15 minutes. If clothing or footwear has been in contact with acetic acid, then they should be removed under a shower. Wash any contaminated clothing or footwear before next use. If any irritation, pain or signs of toxicity occurs, seek immediate medical attention.

If splashed into an eye, the eye should be washed with copious amounts of water for at least 15 minutes. The eyelid should be held open during this procedure to ensure all parts of the eye are washed. Following this, seek immediate medical attention.

If ingested, rinse mouth with water, do not induce vomiting. If victim is unconscious, do not attempt mouth-to-mouth resuscitation. Take to hospital to get immediate medical attention.

General safety

All the chemicals used in the preparation of biodiesel & detailed here are extremely dangerous. Care & respect of these chemicals should be taken at all times. This will include ensuring that they are handled with caution, stored safely & disposed of appropriately. Some of these chemicals also have a limited shelf life, therefore it is important that only enough of these chemicals that you will use making each batch of biodiesel should be ordered & stored on site.

The same principle should be adopted for all these chemicals. That is - it is important that only enough of these chemicals that you will use making each batch of biodiesel should be ordered & stored on site. This is because it reduces the risk of storing & handling dangerous chemicals. Also, if biodiesel is being made at your home, it is doubtful that your household insurance will include cover for storing large amounts of volatile & dangerous chemicals. Your neighbours may also take a negative view to this too.

General chemical advice

If you do purchase more chemicals than you need at any one time, ensure that you decant the portions that you intend to use into smaller containers. These should be new containers that are labelled correctly. Also, ensure that there are enough materials at immediately at hand to be able to cope with any spills & a hose pipe with a spray trigger should also be ready.

All first aid equipment that could also be needed should also be available immediately to hand. It is also advisable to have a second person standing by, ready to administer first aid & to oversee the decanting process. If a splash of chemical gets into someone's eyes, they will need someone to help, rather than blindly feeling about for an eye wash or a hose pipe.

The storage of the chemicals should also ideally be in a locked & clearly labelled box. As sodium hydroxide, potassium hydroxide & sulphuric acid all react with steel; therefore a suitable, lockable plastic box should be used. Wherever you source your chemicals from should be able to advise & provide a suitable store.

Never keep these chemicals in the house, or in reach of children or animals.

Oil advice

The chemicals are not the only hazard. Cooking oil can cause a slip hazard if any spills on the floor. It could even cause something to slip from your hand. Handling things with oil covered gloves will always prove to be problematic. All items that could become slippery should therefore be continually wiped down after they are handled to keep them free of oil contamination. It would also be advisable to cover the immediate area where you are working with sawdust, or oil absorbing granules like kitty litter.

All items used to clean spills & runs should be placed into a suitable bin at the earliest opportunity. The bin & its contents should never be mixed with any household rubbish & should be disposed of safely & quickly at a suitable location. This could be at the local refuse tip where they should have facilities to cope with chemical waste. If this is not the case, make enquiries with your local council regarding the safe disposal of the waste.

Also, in regard to bins. Anything that has been soaked in oil or chemicals run the risk of spontaneous combustion. This is not a joke, fires are started on a regular basis due to the spontaneous combustion of oil soaked material. If oil is left on rags, the oil will react to the air in a process called oxidation, which will cause heat. The rags will also act as insulation & therefore the heat will rise. It can happen with any material such as cardboard & sawdust. Putting oily rags in the washing machine can also increase the heat, as does leaving them in a pile, in the sun.

To reduce the risk of spontaneous combustion, anything contaminated with oil should be placed in a metal can with an airtight metal lid until you can wash them by hand & dry on a line. If they do need to be left out, either leave them spread out on an external concrete floor so that they can not act as insulation against each other, or place them in a bucket of water. Following these steps could avoid a potential fire hazard.

Heat advice

When making biodiesel, the application of heat is used. This involves heating the cooking oil. If a heating ring with an open flame is used, such as with a gas burner, then this will pose a serious risk of fire as methanol is highly flammable & burns with an invisible flame, which means any potential spills & splashes could ignite & be unnoticed.

Also, the fumes from the methanol could readily ignite in the presence of an open flame. Also, cooking oil carries the risk of fire. Therefore, it is imperative that only hotplates or electric rings are used in this process.

It is also important never to leave anything unattended during the heating process. The addition of a heat alarm would therefore be beneficial.

Electrical advice

Electricity can cause sparks. Therefore it is important that only electrical items are used which are safe for use. An electrical spark in the presence of methanol could cause an invisible fire. Electrical motors (such as those found in drills) also cause sparks, therefore electrical drills should never be used to stir the cooking oil.

The use of an RCD is also advisable so as to protect the area & the individuals from the chance of electrocution due to electrical faults. Electrical cables would also pose a trip hazard if trailed across the floor, therefore it is advisable to secure the cables at height as much as possible.

PPE advice

Wherever you look, there will be advice on which PPE is best, but it is a complicated subject area, therefore it is always best to err on the side of caution.

It was previously stated that a full face mask would be suitable for use with methanol. It can be suitable, but should be thought of as a minimum standard.

Ideally, a powered air-purifying respirator (PAPR) or a full face mask should be used. However, any respirator/mask is only as good as the installed filter.

For instance, a filter suitable for use with the 3M 6000 reusable full face mask (model 6700) would be the 3M gas & vapour particulate filter, 6000 series 6098 AXP3R. This filter would also be suitable for use with the 6000 reusable full face mask model 6800, model 6900 & 7000 series 7907S. The mask is reusable, but the filter is not.

Any mask or respirator manufacturer should be able to tell you which of their products would be suitable to be used with methanol. 3M have a website that will do this[45].

Coveralls should also be used. It would be best to use a 'chemical, nuclear, biological' suit. These can be obtained cheaply from army surplus stores. Gloves or gauntlets should be chemical resistant & can be bought cheaply online. It is also advisable to obtain chemical resistant Wellington boots, although these are not so cheap.

Other sources

Wherever you decide to purchase the chemicals & PPE from, it is important that you select a reputable company from your country, rather than opting for the cheapest Chinese product you can find on the internet.

---

[45] https://sls.3m.com/ - 06/07/2020

It is also advisable to communicate to the supplier what your intended use will be. For instance will be of no use buying Wellington boots that do not have oil resistant soles, or try to buy 2 litres of methanol from a company that only deals in multiples of 1,000 litres, or the wrong grade sulphuric acid.

You will not be the first person to contact them with requests regarding biodiesel manufacture at home & they should know exactly what you need, rather than what you may think you'll need.

Also, DIY biodiesel has been in the news due to accidents[46]. It may be interesting to look at some other examples where individuals have not followed basic safety tips & have gone on to regret their actions. There is one example of a man in Northampton, UK who 'blew up his garage' making biodiesel[47].

All cases like these should be treated as how not to make biodiesel. There is a long list of incidents on the website of biodiesel magazine[48] & on the make-biodiesel.org website[49]. Despite many being industrial accidents, they could be interesting reading to get some more background information.

Other chemicals

If you stick to the basic biodiesel instructions in this book, the chemicals listed in this chapter are the only chemicals that you will need.

---

[46]   https://www.spokesman.com/stories/2009/apr/28/dangers-accompany-home-biodiesel-production/ - 06/07/2020

[47]   https://www.telegraph.co.uk/news/newstopics/howaboutthat/2463557/Man-blows-up-garage-trying-to-make-biodiesel.html - 06/07/2020

[48] www.biodieselmagazine.com/articles/4055/biodiesel-plant-safety/ - 06/07/2020

[49] http://www.make-biodiesel.org/Biodiesel-Safety-list/ - 06/07/2020

Therefore the information has been provided, but if you experiment with other biodiesel recipes that you may find on the internet, you will need to look up the safety information for those chemicals & familiarise yourself with them before you attempt any work.

Chapter 8 – Equipment

Apart from PPE, there will be some other equipment that will be needed. Some of this equipment will be used infrequently & some all the time. Also, if your batches start small, then increase as you ramp up production, you will need bigger & different equipment. For instance, it will be possible to stir a 1 litre batch by hand, but not a 200 litre drum, therefore different tools will be required for the same job.

Also, it may be tempting to use equipment that you may find in your kitchen, but NEVER USE YOUR KITCHEN EQUIPMENT. Any equipment you use in the preparation of biodiesel will be contaminated with dangerous chemicals. For instance, if you use a wooden spoon from your kitchen, then the chances are that these chemicals will find their way into your food the next time the spoon is used in the kitchen. Therefore, if you use it for the preparation of biodiesel, label it, keep it safe & lock it away outside the house & away from any children or pets. It may also be a good idea to replace anything you've liberated from the kitchen before its absence is noticed.

Safety equipment

All the safety equipment should be obtained before attempting any scale of biodiesel production. There is as much chance of injuring yourself making 1 litre as 200 litres.

Eye wash

The first element on the list is eye wash. This can usually be sourced from numerous places, online retailers, commercial suppliers of industrial equipment, safety retailers & even DIY shops.

Eye wash is usually sold in either small single use plastic ampoules that contain boiled water, or larger sized bottles.

It may be advantageous to obtain an eye wash station which can then be wall mounted next to a first aid kit. That way, the safety equipment can all be kept in one location.

First aid kit

It is also advisable to obtain a suitable first aid kit. The first aid kit will not contain everything that you need for first aid, but it is used for immediate use for small injuries. Again, a wall mounted unit would be advisable so that it can be mounted next to all the other safety equipment.

These can also be obtained from many places including supermarkets, chemists, online retailers, commercial suppliers of industrial equipment, safety retailers & even DIY shops.

Fire extinguisher

At least one suitable fire extinguisher should be sourced & located with the other safety equipment. An extinguisher suitable for use with methanol should be obtained. This would be either powder or foam, or both. This is because they operate using different principles.

Dry powder extinguishers work by cooling the flames & can also interfere chemically with the burning reactions. However, it does not cool or seal anything that is on fire. It is excellent to damp down a fire if you want to get past it, but will have limited use if the fire reignites.

There are different types of foam extinguishers. Most can be 95% water with a frothing agent (just like soap bubbles) & also a chemical to make the froth stick together. This is designed to float on top of the fire, but as methanol is an alcohol, it mixes readily with the water & therefore dilutes the foam, but as water will dilute methanol, if you have enough, it will eventually put out a methanol fire, but you will need far more than one fire extinguisher.

There are some foams available that are alcohol tolerant & these will work better than the usual foam extinguishers.

Standard AFFF foam extinguishers are the water type, therefore a foam extinguisher that does not contain AFFF foam should be obtained. These contain alcohol resistant foam (AR foam) & are more expensive than the AFFF type.

Foam does also cool the fire so that the chance of re-ignition is reduced, but a six litre fire extinguisher may only operate for up to 30 seconds, therefore it may be advantageous to obtain more than one.

The fire service will use water to tackle a fire. Water will slowly dilute the methanol & they will use a hose nozzle that will spray a fine mist of water. This will have the added effect of cooling the area too.

There are also carbon dioxide (CO_2) extinguishers on the market which would prove useful indoors as they will remove the oxygen & without oxygen a fire can not stay alight, however, outdoors they will be ineffective.

Therefore, a fire extinguisher will only be useful for a limited time in limited circumstances; therefore it would be prudent to have a selection at hand, along with a hose pipe with a shower head which should then be used to continue with after the fire extinguishers have been exhausted.

Automatic fire extinguishers are available online. These can be installed over any potential source of fire. These systems consist of self contained water filled canisters that have a sprinkler installed at the bottom. They operate just as any other sprinkler. The sprinkler has an alcohol filled glass bead fitted & will break when subjected to the heat from a fire.

When the bead breaks, the water is released because the bead was blocking the outlet. The glass beads are colour coded to indicate at which temperature they are designed to break. Orange (57°C 135°F), red (68°C 155°F), yellow (79°C 175°F), green (93°C 200°F), blue (107°C 225°F), purple (149°C 300°F), black (191°C 375°F), black (246°C 475°F) & black (329°C 625°F).

Fire blanket

There are numerous fire blankets available on the market & any of these will be a useful addition to the safety equipment. They can be used as a shield to anyone needing to get past a fire & can also be used to smother flames. Providing a blanket is labelled to be compliant with EN1869, it will reach the minimum standard needed. These can also be of more impact on a fire than a fire extinguisher for those who do not know how to use a fire extinguisher effectively. They are useful in fires that reach up to 550°C (1,022°F). If a higher rating is required, then welding blankets can be used as they are designed to be used in temperatures up to 980°C (1,800°F).

A fire blanket will always be a useful addition. It can be used for smothering small equipment fires, a fire shield to help an individual move past a fire & to smother flames on a clothing fire.

Again, as before, these can also be obtained from numerous places including chemists, online retailers, commercial suppliers of industrial equipment, safety retailers & DIY shops.

Shower hose

A water filled hose pipe with a trigger operated shower head should be kept ready & at hand at all times. It will be useful for tackling methanol fires as the water will dilute the alcohol, it will cool & put out a fire if an individual has a clothing fire. Water is cheap & there should be a ready supply at hand.

However, the water tap must be switched on first so that it is ready to use if needed. Valuable time will be lost if you first have to run to the tap to switch it on before you can use the hose. It could mean the difference between life & death, or a small fire & a massive fire.

Spill kit

It is always important to have a spill kit readily available. This can be a bag of sawdust, sand, kitty litter, babies nappies, earth or even a commercially available spill kit. Which type is used is not important. What is important is that there is enough to soak up any potential spills that may occur. It is also important that these materials are disposed of appropriately & safely if they have been used. Do not place any of these items in the domestic rubbish after they have been used.

Personal protective equipment (PPE)

There are several items of PPE that you should have before making any biodiesel. The first of these is gloves or gauntlets.

Gloves & gauntlets

Any gloves or gauntlets that you use must be suitable for chemical & methanol use. There are universal standards that help determine the suitability of the gloves. EN388 defines gloves due to their abrasion resistance, cut resistance, tearing strength, puncture resistance, cut resistance (to ISO13997 standards) & impact protection.

Abrasion resistance is defined by a number, 1 – 4. The higher the number, the better the glove will stand up to abrasion.

Cut resistance (coup test) is given a number 1 – 5, where the higher the number, the more it will stand up to being cut with a knife. If the glove dulls the knife, then cut test ISO13997 is used instead.

Tearing strength is defined by a number 1 – 4 & as before, the higher the number, the better it will resist tearing. Puncturing is also measured by a number 1 – 4 & the higher the number, the better the material stands up to being punctured.

Cut resistance, TDM test ISO13997 is the test used if the material has dulled a knife. The test is defined by a letter A – F, where F is the highest level of protection. It is used in place of the coup test. Impact protection is given as a letter P. If there is no P present, the gloves do not offer any impact protection.

 This information will be displayed on the packaging of the gloves/gauntlets as a small pictogram.

The numbers & letters under the EN388 are written in the order defined above.

Therefore, the above pictogram will indicate the following: The gloves have poor abrasion resistance (only 1 out of 4), the cut resistance is only slightly better (only 2 out of 5), the tearing strength is OK (3 out of 4) & the puncturing resistance is very good (4 out of 4). Cut resistance to ISO13997 only scores a poor B (A – F, where F is the highest) & finally a P is listed, therefore it offers impact protection.

Chemical resistance for gloves is covered under EN374, therefore all gloves must comply to this standard to be considered to be chemically resistant. This is split into two tests. EN374 – 2 tests penetration. This is how the glove stands against the movement of a liquid through the glove (including its imperfections) at a non-molecular level. EN374 – 3 tests permeation. This is the process by which a chemical moves through the glove at a molecular level.

 A glove does not need to pass both tests to pass EN374, but if they do not, the gloves will only protect against certain substances. Therefore any glove or gauntlet that passes EN374 – 2 but not EN374 – 3 will display this symbol.

 Any glove that passes EN374 – 2 will display this symbol. If it obtains level 1 grade, it is waterproof. If it achieves a level 2 or 3 grade, then the glove is also resistant to micro-organisms.

EN374 – 3 measures how long a glove or gauntlet can withstand a chemical[50] before it permeates through the glove. Level 1 must withstand more than 10 minutes, level 2 more than 30 minutes, level 3 more than 60 minutes, level 4 more than 120 minutes, level 5 more than 240 minutes & level 6 more than 480 minutes.

 If a glove or gauntlet achieves a minimum of level 2, then it can display this symbol. Therefore only gloves or gauntlets that display this symbol will be suitable for use in the production of biodiesel.

Any gloves listed as medical examination gloves, or similar will not have the requisite chemical safety requirements for biodiesel use, therefore they should not be used under any circumstances.

---

[50] One of three chemicals chosen at random from: Methanol acetone, acetonitrile dichloromethane, carbon disulphide, toluene, diethylamine, tetrahydrofuran, ethyl acetate, heptane, sodium hydroxide & sulphuric acid (96%).

Coveralls

Appropriate coveralls should also be used. Coveralls should be used in conjunction with a protective hood, over sleeves. These are available in different grades along with some being flame retardant.

EN14605 covers standards for chemical protection. As was seen with the gloves, the level of protection is given a number.

 Type 1 is gas tight protective clothing which gives protection against solid, liquid & gaseous chemicals (see also EN943 – 2). It is represented with this symbol.

 Type 3 is liquid tight protective clothing which offers protection against a strong directed jet of liquid (see also EN14605). It is represented with this symbol.

 Type 4 is liquid tight protective clothing which offers protection against a liquid that is not under pressure (see also EN13034). It is represented with this symbol.

 Type 5 is particle tight protective clothing that offers protection against solid airborne particles (see also ISO139872-1). It is represented with this symbol.

 Type 6 is partially spray tight protective clothing which offers protection against a liquid mist (see also EN14605). It is represented with this symbol.

It should be clear that types 1 & 2 offer the highest level of protection & are designed for working in highly hostile environments.

They are designed to incorporate a breathing apparatus or an air line. Types 3 & 4 are made from non-breathable, chemically resistant fabrics & have sealed seems. Type 3 is designed to protect against a jet of liquid, whilst type 4 is designed to protect against a high pressure spray. Types 5 & 6 offer little or no protection in biodiesel production.

Due to the potential of methanol catching fire, the overall should also be flame resistant. ISO11611 & ISO11612 covers garments that protect against heat transmission as well as against flame spread, whilst ISO14116 (which superseded EN533) covers the resistance to the spread of flame, which will be more appropriate to in biodiesel production. Coveralls which meet ISO14116 are designed to protect against occasional brief contact with small flames. They fall into one of three types.

Index 1 are designed so that the flame does not spread, there is no flaming debris, no after glow, but a hole may be formed.

Index 2 are designed so that the flame does not spread, there is no flaming debris, no after glow & no hole may be formed.

Index 3 are designed so that the flame does not spread, there is no flaming debris, no after glow, no hole may be formed & the after flame must extinguish in less than 2 seconds.

Clearly, index 3 garments will offer the best level of protection, but no pictograms are used for this standard. It would be advisable to also use sleeve covers along with a hood. The material standards for these will be as for coveralls.

It may be advantageous to check army surplus stores for 'nuclear, chemical & biological' (NCB) suits. These provide inexpensive protection, but if you do buy a NCB suit, ensure that it is new, not used.

Footwear

Chemical resistant footwear should also be used as any spills will end up on the floor & therefore protection must be provided to stop any injuries occurring to anyone who inadvertently stands in any chemicals. The Wellington boot type would offer the best protection as it will also protect the user from splashes too.

With chemical resistant footwear, the level of protection is also determined by the amount of time it will stand up against the ingress of chemicals. EN13832 – 3 gives the following five levels.

Level 1 allows the ingress of chemicals between 121 – 240 minutes.

Level 2 allows the ingress of chemicals between 241 – 480 minutes.

Level 3 allows the ingress of chemicals between 481 – 1,440 minutes.

Level 4 allows the ingress of chemicals between 1,441 – 1,920 minutes.

Level 5 allows the ingress of chemicals anytime after 1,921 minutes.

Unfortunately, this chemical resistance level does not specify to which chemical it applies, therefore it could be for a chemical that is not applicable for biodiesel production, therefore EN943 covers chemical protection & is again graded as numbers.

Type 1 offers gas tight protection, as well as protection from solids, liquids & gasses (EN943).

Type 1a offers gas tight protection & is to be used inside the suit with breathing apparatus (EN943 – 1).

Type 1a-ET is the same as 1a, but for use by emergency teams (EN943 – 2).

Type 1b is for gas tight protection with a breathing apparatus used outside the suit (EN943 – 1).

Type 1b-ET is the same as 1b, but for use by emergency teams (EN943 – 2).

Type 1c offers gas tight protection & is to be used with an air fed suit, not with breathing apparatus (EN943 – 1).

Type 2 is to be used with an air fed suit that is not gas proof (EN943 – 1).

Type 3 is liquid proof (EN14605).

Type 4 is spay proof (EN14605).

Type 5 offers protection from particles (ISO13982 – 1).

Type 6 is for low risk exposure from liquids (EN13034).

It should therefore be clear that only types 1, 1a & 1b will be suitable for biodiesel production. The correct types are known as HAZMAX boots.

Also, ISO20344 sets out certain standards for safety footwear that are relevant. That is their slip resistance. There are three levels.

SRA – tested on a ceramic tile wetted with a dilute soap solution.

SRB – tested on smooth steel with glycerol.

SRC – tested under both conditions.

Therefore suitable Wellington boots will also need to have a slip resistance rating of SRB, or SRC.

Goggles or face protection

In some situations a pair of goggles may be suitable, but for others, a face shield may be more favourable. Both these, along with sunglasses, visors & safety spectacles are all covered by EN166. The standard covers various points, but not chemical resistance, however, in the USA, ANSI Z 87.1 states that chemical resistant splash goggles must fit snugly around the face & have hooded vents to stop the ingress of any liquids but still allow for ventilation.

Goggles can always be used in conjunction with a face shield, but the use of a face shield without goggles should never be undertaken.

It should be noted that in the UK, eye & face protection in hazardous areas is a requirement under Regulation 4 of the *Personal Protective Equipment at Work Regulations 1992*. Although biodiesel production is likely to be undertaken at home, this regulation should still be followed. It is always advisable to follow safety measures at all times.

Enhancements to face shields that would also be beneficial would be to always opt for models that have chin guards (to stop splashes travelling upwards towards the face) & also side shields (to protect the side of the face).

In the UK, the Health & Safety Executive (HSE) categorises between the different masks that are available on the market. They can be 'respiratory protection equipment' (RPE) which filters the air to remove harmful substances, or, 'breathing apparatus' (BA) which provide clean air. RPE can be powered, or non-powered & BA are fed from an external powered source. Either of these can be tight fitting, so as to make a good seal with a persons face, or loose fitting such as with hoods & helmets.

The HSE state that when selecting RPE or BA, it must be adequate & suitable; therefore you must know what it should be protecting you from.

In the case of biodiesel production, it would be methanol (CH_3OH) vapours. If the concentration of these vapours are high, then BA should be used, not RPE.

However, it is assumed that all biodiesel production will be undertaken outdoors, therefore RPE will offer adequate protection, provided you use the correct filter. That is one that is designed to deal with methanol (CH_3OH) vapours, not one that is designed to deal with particles. It is worth remembering that particle filters do not protect against any vapours.

Also, only powered masks, powered helmets & air hose fed masks are designed for continual use for more than an hour. Everything else is designed to only be used for less than an hour. Also, if an individual wears glasses, they are not compatible with a full face mask as the arms will break the seal. The same is true if an individual has facial moles, stubble or a beard, therefore a helmet design would need to be used.

Also, as working with methanol (CH_3OH) vapours can cause a potentially flammable or explosive atmosphere, it would be advisable to use alloy free, antistatic RPE.

Relevant standards which may be of further reading, depending on which type of RPE you select would be. Power assisted devices:

EN12941, EN1835 & EN270 – Power assisted filtering devices incorporating helmets or hoods. This categorises three classes for the equipment, TH1, TH2 & TH3. The filters are designated TH1P, TH2P & TH3P.

EN12942 – Continuous flow compressed air-line breathing apparatus. There are four light duty grades, 1A, 2A, 3A & 4A. There are also four heavy duty grades, 1B, 2B, 3B & 4B.

Self compressed air devices:

EN137 – Self contained open-circuit compressed air breathing apparatus.

EN145 – Self contained close-circuit breathing apparatus compressed oxygen or compressed oxygen-nitrogen type.

EN1146 – Self contained open-circuit compressed air breathing apparatus incorporating hoods. TH1P, TH2P & TH3P.

EN402 – Self contained open-circuit compressed.

In the last chapter, a suitable filter for use with methanol was noted as 3M gas & vapour particulate filter, 6000 series 6098 AXP3R, however, this is a single use filter which is suitable for only 40 minutes use when the methanol vapour concentration is less than 100 parts per million & only 20 minutes if the vapour concentration is less than 500 parts per million. If the vapour concentration exceeds this level, then BA should be used. To ensure maximum use from one of these filters, individuals should never stand over open canisters of methanol; it should be pumped from container to container wherever possible & all activities undertaken outside, never indoors.

Misc safety equipment

There are other items that could prove to be useful whilst making biodiesel.

Mobile phone

It is always important to have a mobile phone to hand. If an ambulance or the fire brigade need to be called, then every second will count. If it takes two minutes to run to a phone, then that will be an extra two minutes someone could be waiting for medical attention, or an extra two minutes that a fire has been allowed to spread before the fire brigade arrives. Saving time could save lives or property, therefore a mobile phone is imperative.

Thermal imaging camera

A worthwhile investment would be either a camera with a thermal imaging lens, or a hand held thermal imaging gun. This measures the temperature, which is important as methanol burns with an invisible flame; therefore it will be obvious to anyone who is using the camera if a fire has occurred, even if it is not immediately apparent to anyone else.

Figure 39 Thermal image of a pot on stove (Public Domain © 2020)

There are complete camera's available, stand alone hand held devices that look like a hand held hair dryer & even those which attach to a smart phone.

Anyone using the thermal imager must be trained beforehand, so that they are familiar with the readout, although they are somewhat basic & simple to use. The readout displays the temperature of the hottest object as a picture or video, with the hottest part being white or red.

When using a thermal imaging camera, it will be apparent to the operator if there is a fire. The previous image shows a cooking pot on a stove top. Note the bottom of the pot where the heating element is located, this is shown as being white hot.

A N Other

Clearly it will be beneficial to have an individual standing by during all stages of biodiesel production. This individual should be there to act purely as a safety coordinator. Ideally this individual will be equipped with a hose pipe with a shower head, a thermal imaging camera & a mobile phone.

This individual should be fully briefed with the operation & will be standing by ready to administer first aid or to call for assistance if needed. If an individual who is making biodiesel splashes methanol on themselves or gets it in their eye, having another individual available to offer immediate assistance could save a life.

Vinegar

Vinegar neutralises potassium hydroxide (KOH), so it is worth keeping this to hand in a squeezable bottle to neutralise any spills or splashes that may occur. An old washing up liquid bottle may be a suitable receptacle as it will allow for a controlled squirt or pouring of the vinegar. It is however important that it is correctly labelled as vinegar & marked to stop it being used in the kitchen.

Laboratory & useful equipment

There will also be a number of items that will be required for the preparation of biodiesel. Not all items will be needed for small batches, but if the production level increases, then more professional equipment will be required.

Hand transfer pump

The use of a hand pump to transfer dangerous chemicals from one receptacle to another will always be far safer than pouring the chemical from one container to another.

A small siphon pump will be needed to measure out small quantities of methanol. Small fuel transfer type hand pumps can be obtained online for less than £20 pounds & if a hand siphon pump is to be used for more than one chemical, then one siphon should be used for each individual chemical. Label each of the pumps so that cross contamination does not occur.

Weighing scales

A small set of laboratory scales will be beneficial. One that can weigh 0.1g would be best for weighing small measures. These can be bought online for less than £10. Also, a set of kitchen scales would also be of great use which can be used for the heavier measures. Again, these should not be taken from the kitchen, nor used in the kitchen once they have been used in biodiesel production. They should therefore be labelled to ensure that they are no longer used for food & to ensure they are never used in the kitchen. Digital scales may be easier to read than traditional scales.

Blender

A blender with a glass container would be useful to mix ingredients, but it should have a glass jug, not plastic. Also, as these have the blades inside the jug, there are usually several washers fitted to the spinning mechanism. Biodiesel will eat into any seal that is rubber, therefore before the blender is first used, the blade & sprocket assembly should be disassembled & all rubber washers replaced with Viton washers, or any other biodiesel safe material.

Also, a blender should not be taken from the kitchen, nor used in the kitchen once they have been used in biodiesel production. It should therefore be labelled to ensure that it is no longer used for food & to ensure it is never again used in a kitchen.

It may be possible to obtain a laboratory blender, but the price could prove to be prohibitive & it is unlikely that one would be found in a second hand shop or in an online auction.

Pasteur pipette aka eye dropper (graduated)

There are two types of Pasteur pipettes available. The first type is disposable & is made from polyethylene. These are not suitable for use in biodiesel production as they can break down with the use of strong chemicals. The second type are made from borosilicate glass & have a rubber bulb at the top.

These are an excellent way for transferring small amounts of liquid. However, never use the pipette without the rubber bulb & never over fill the pipette so that the bulb comes into contact with any chemicals.

These usually come in 1, 2, 3 & 5ml sizes, so it would be best to have a small selection.

Hydrometer

Biodiesel has a density of 0.86 – 0.90g/ml (water is 1g/ml). As you may wish to check the density of any biodiesel that has been produced, a hydrometer will be needed for this task.

Typically, most on the market are for brewers & they are designed for 0.990 – 1.070g/ml, therefore a hydrometer that operates within the range of 0.85 – 0.90g/ml is needed. BS718 M50SP is the standard for hydrometers in this range & are specifically designed for testing oil & petroleum products. These can be obtained for approximately £20. Another type that can be used is a hydrometer built to BS718 M100 which operates within the range of 0.800 – 0.900g/ml. These are not petrochemical specific, so are a little cheaper.

When determining the reading of the hydrometer, it is important to check the level of the liquid, not the top of the meniscus (this is when the surface of the liquid forms a concave edge). The lowest part of the meniscus should be read, not the top part of the meniscus.

The hydrometer will also need to be used in conjunction with a thermometer.

Thermometer

The use of a thermometer will be advantageous in biodiesel production. Either an alcohol filled borosilicate type or an infrared gun type will be suitable. These can also be obtained for less than £20. A jam thermometer will cover all the temperatures needed for biodiesel production.

Borosilicate laboratory beakers

There are currently two types of beakers available. The first type is made from polyethelene.

These are not suitable for use in biodiesel production as they can break down with the use of strong chemicals. The second type are made from borosilicate glass & these are suitable.

Glass beakers with pouring spouts should be obtained in various sizes from 10 – 500ml sizes. It may be useful to obtain these sizes in various shapes – Erlenmeyer flask (aka conical flask), cylindrical beaker & graduated cylinder (aka measuring cylinder).

These should never be used in a kitchen if they have been used in biodiesel production. They should therefore be labelled to ensure that they are not used for food preparation or used in a kitchen.

Borosilicate Petri dishes

These are also known as Petri plates or cell-culture dishes. They are excellent for receptacles for placing & weighing small amounts of liquid or solid chemicals.

Borosilicate stirring rods

These are borosilicate glass rods between 10 – 40cm used to stir liquid chemicals.

Electronic pH meter, litmus paper & phenolphthalein

The pH can be measured with either litmus paper or with an electronic pH meter. The paper is usually sold in booklets which have the scale printed on the book cover. They operate by just dipping a paper strip into the liquid. The paper then changes colour & it is compared to the scale on the book.

Alternatively, an electronic pH meter can be used which will give the reading as a numeric value. These can be very accurate & can be purchased for less than £10.

Small quantities of phenolphthalein can be sourced online for less than £10.

Laboratory test sieves

It will always be worth having good quality stainless steel laboratory test sieves. These are available in a variety of sizes typically between 125mm down to 20 microns, although the larger sizes will be of no use in biodiesel production. The whole point of using these sieves is to remove unwanted particles from WVO, therefore 10 – 5mm would be the largest size that would be needed, then work downwards, towards the smallest size.

These sieves are not cheap. They can cost upwards of £100 per sieve, therefore it will be worth checking internet auction sites regularly for suitable sieves. These sieves stack on top of each other with the widest meshed sieve at the top & smallest meshed sieve at the bottom. The WVO is then slowly poured in at the top & as the oil works its way down through the stack of sieves, any solid particles will be caught in the sieves & filtered out. This should leave pure WVO to exit at the bottom as all the particulate matter has now been removed.

Electric hotplate

An electric hotplate can be used to bring the oil up to temperature. Never use a naked flame to do this as it could cause the methanol to ignite. These can be obtained online for less than £20, but will only be suitable for the creation of very small batches.

Plastic home brew buckets

These large 25 litre buckets can be bought very cheaply in any shop that sells home brew equipment for brewing alcohol. They can be between 20 – 30 litres. It does not matter what size is obtained, but the larger they are, the heavier they will be to lift. These can be bought for approximately £15 each.

Some are available that have a spigot tap fitted at the base. This type is useful as it will allow the removal of chemicals that have settled at the base of the bucket. These are similarly priced as those without the tap fitted. At least two of each type would be useful.

Stainless steel cooking pot

A large cooking vessel will be needed to warm the oil. A large stock pot will be needed for this process, typically one that can hold 5 – 10 litres would be best. It may be worth getting a stainless steel maslin pan. These are used for making jam & usually 9 litres in size & have graduations marked on the inside. They also have a pouring spout & large handle, which make them more useable than stock pots.

Pen & notepad

It will be important to keep notes on the biodiesel production process at various stages along the way, therefore a pen & notepad will be needed. Not only to refer back to if the biodiesel has not worked out, but also, it will be advantageous to keep a record of all the data that is created because studying the data at a later date will add to useable knowledge which will then streamline the process for any experienced biodiesel brewer. This may sound like a waste of time, but it is one of the most important habits to adopt. It may mean the difference of a 30% success rate or a 95% success rate when making the biodiesel.

Small lab spoons

These will be invaluable when measuring very small quantities of dry or wet chemicals. They are sometimes called micro spoons, scoops or spatulas. They should always be made from stainless steel & a set can be purchased online for less than £20.

Preserving jars

It may be advantageous to obtain some preserving jars. The kilner type jars would be ideal as a small sample may need to be kept, or a small sample may need to be experimented or tested, therefore a strong, airtight container will be needed. The kilner type jars with the clip top are ideal as they are strong & airtight & also have the advantage of being glass, therefore test samples can be visually checked without opening the jar. A few 500ml jars should be enough. A six pack of 500ml jars will cost around £20.

Brushless drill

Small mixes can be stirred by hand, but larger quantities of biodiesel will need to be stirred vigorously for prolonged periods. Stirring for any length of time can be both dangerous due to the methanol fumes & the danger from chemical burns due to the corrosive chemicals. It is therefore far better to use an electric drill to undertake the long & dangerous stirring that will be needed, but as electric drills spark, the only option is a brushless drill. These can be very expensive, but can be obtained for as little as £50.

Mixing paddle

If a drill is to be used to mix the biodiesel, then a mixing paddle will also be needed. These are drill attachments that tend to be used for mixing plaster or cement screeds, but are equally at home in biodiesel production. These can be obtained at DIY stores or online for approximately £15.

300ml tin

The best way to test for viscosity is to use a tin (approximately 300ml or larger) that has a small hole drilled in the side, near the base. Any liquid that is placed in the tin will flow through the hole & this can be timed. A viscous liquid will take longer than a less viscous liquid to escape. This method will not be able to give a scientific reading (viscosity is expressed as centipoise (cP)), but it will allow a determination as to whether a liquid has been made less viscous providing it has been timed accurately.

There is also a scientific instrument called a Zahn cup viscosimeter available that will perform exactly the same task. These can cost upwards of £30 & the reading it gives is then looked up on a chart to see the resultant centipoise. This reading will mean nothing to the average man, but the hole in the tin is easy to understand & accurate.

Stopwatch or watch with a second hand

To accurately gauge how viscous a liquid is, accurate timing is vitally important, therefore a stopwatch or a watch with a second hand will be ideal. If a suitable watch can not be found at home, a stopwatch can be bought online for less than £10.

Ladle

A ladle may be needed to transfer small quantities of liquids from one jar, or bucket to another. It is therefore worth keeping one to hand. A stainless steel ladle would be the best choice & they can be bought online for less than £10.

In-line fuel filters

In-line fuel filters will be useful for filtering your fuel. It is best to purchase these in packs of ten as they will be far cheaper this way as online a 10 pack can be bought for less than £20, but a single filter costs around £3. These filters have an opening at each end. One is the inlet, one is the outlet. Hence the name in-line. They are designed to have tubing attached at each end, therefore they have push fittings for attaching rubber type hoses. If the filter has 6mm fittings, then it will need 6mm hoses, if 8mm, then 8mm hoses will be required.

2m fuel hose

If filtering is to be undertaken using the in-line filter, then a fuel hose will be needed to attach to each end of the filter. A 2m length cut in to two should be long enough for most jobs. It is however vitally important that a biodiesel compatible fuel hose is bought. This could be Viton or any fluroelastomer fuel hose. A 2m length should cost no more than £10.

Small funnel

Moving the biodiesel through the fuel hose & filter will be impossible without a small funnel that will fit in to the hose. A small stainless steel funnel will be best the best option as it will be resistant to the biodiesel, hard wearing & easy to clean. These can be bought online for less than £10.

Biodiesel storage container

This can be anything from a portable fuel can to an oil drum & which one/s will be dependant on your own particular needs. Storage will be covered in detail later in this book.

Flexi tubs

It will be worth obtaining a few flexi tubs. These are large flexible buckets & are perfect for short term storage of the by-products made during the biodiesel process. They are cheap & very stable. They can be bought at DIY stores or online & usually be bought for about £5 each. A minimum of two should be bought.

Petrochemical diesel (sample)

When making biodiesel, it is important that it replicates petrochemical diesel as close as possible, therefore having a sample to hand so that the biodiesel can be immediately compared to it is advisable. A 200ml sample should be enough & it should be stored in an airtight container in the dark to stop it degrading. Replace the sample after six months.

Large scale production equipment

If you intend to produce biodiesel in any large quantities, there are some things that will make life a little easier.

Fume cabinet

If large scale production is to be the outcome of the biodiesel manufacture, then it would be worthwhile constructing a fume cabinet. These are somewhat expensive to buy at approximately £2,000. Therefore a DIY version can be built for less than £200. A simple internet search for 'DIY fume cabinet', 'DIY fume cupboard' or similar will reveal many home built designs that would be suitable.

Please ensure that if you do construct a fume cabinet that you fireproof the internal faces & vent the cabinet to the outside.

Chapter 9 – The first batch

Having read this far, it's now time to reward yourself by making the first small experimental batch of biodiesel. Nothing too complicated provided that all the steps are followed.

First to recap. Vegetable oil (SVO) is made from millions of tiny molecules. These molecules are made from glycerol molecules, each of which have three chains of fatty acid methyl esters (FAME's) attached to them. The process of converting SVO or waste vegetable oil (WVO) into biodiesel is just a case of removing the FAME's from the glycerol molecule. These liberated FAME's are put to one side (this is the biodiesel) & the glycerol molecules are left over as a byproduct.

These FAME's do not just drop off the glycerol molecules at will, they are removed by the addition of alcohol because it is far more desirable for the glycerol molecules to attach to alcohol than to remain attached to the FAME's.

As stated in a previous chapter, there are three FAME's attached to each of the glycerol molecules. When one of the FAME's break off, the glycerol molecule still has two FAME's attached, so it is called diglyceride. Break off a further FAME & only one is left attached to the glycerol molecule so it is now called monoglyceride. The presence of diglyceride & monoglyceride in the biodiesel will ultimately damage the engine, so it is important to liberate all three of the FAME's & then separate them from the unwanted glycerol.

That is the whole process of making biodiesel in a nutshell & what will be outlined in this chapter. Making biodiesel is a precise exercise in chemistry, therefore it is important to use an accurate amount of each ingredient. Therefore, being able to test the oil to work out exactly how much of each ingredient to use is vital. Therefore, a test kit will be required to measure the mixture, so first, a test solution will be required.

The test solution

When making biodiesel, the oil will always need to be tested for the presence of FFA's. To do this a titration test solution will be needed beforehand. It is not something that will need to be made continually, as provided the test solution is kept in a dark place in a sealed bottle, the solution should last a long time as only small amounts will be used with each use.

To make the solution, you will need the following:

    1 litre of deionised water
    1 litre glass bottle
    1g of potassium hydroxide (KOH)

The above ingredients must measure exactly 1 litre of deionised water & exactly 1g of potassium hydroxide (KOH).

First remove any label that is on the empty 1 litre bottle & replace with a label that reads 'Potassium hydroxide (KOH) standard solution'.

Pour exactly 1g of potassium hydroxide (KOH) into the deionised water & either stir or shake to allow the potassium hydroxide (KOH) to completely dissolve into the deionised water. Pour the mixture into the empty bottle.

Screw the cap back onto the bottle securely & put the correctly labelled KOH standard solution to one side for later use. It is important to remember that even when diluted in this small amount, the potassium hydroxide will **still be caustic enough to cause burns**.

Test batch №1 – SVO

As this is the first batch & can be thought of as a test batch, or a practice batch, this will be for 1 litre biodiesel, made from SVO.

The equipment

The following items will need to be sourced before starting this first batch.

- Eye wash
- First aid kit
- Fire blanket
- Fire extinguisher
- Trigger operated shower hose on a hose pipe
- Spill kit – this can be a large bag of sawdust
- Gloves or gauntlets suitable for biodiesel production
- Coveralls suitable for biodiesel production
- Wellington boots suitable for biodiesel production
- Safety goggles or face protection suitable for biodiesel production
- Mobile phone
- A second person to act as an observer
- Vinegar in a squeezable bottle
- Weighing scales that can measure 0.1g
- Micro spoons
- Thermometer
- Petroleum hydrometer
- 500ml borosilicate beaker
- 500ml kilner jar
- Borosilicate stirrer
- Electric hotplate
- Electric hand drill with brushless motor
- Mixing paddle to fit electric drill
- Two plastic home brew buckets
- Stainless steel cooking pot (or maslin pan)
- 1 litre of SVO
- Potassium hydroxide (KOH)
- Phenolphthalein ($C_{20}H_{14}O_4$)
- KOH Standard solution

- Pen & notepad
- 200ml methanol (CH_3OH)
- 300ml tin or jug that has a 3 – 5mm hole drilled in the side wall, at the base
- Stopwatch
- 200ml of petrochemical diesel to act as a sample

Step 1 – Safety first & de-watering

After ensuring that all the safety items are ready & that you know how to use them. The next step is to wear the PPE & have all the other safety equipment standing by outside ready for use. Also, place all the other equipment outside ready to use, where you intend to undertake the biodiesel process.

Following this, the first step will be to filter the oil. If it is new shop bought oil that is unused & has been obtained for this experiment, then you should not need to filter the oil, but if it is used waste vegetable oil (WVO), then you will need to sieve out any solids from the oil.

The new (or filtered) oil should then be placed into a stainless steel cooking pan. A jam making pan is ideal (this is called a maslin pan). The maslin pan is wide at the top & narrower at the bottom. This helps with the evaporation process. If you do not have access to a maslin pan, any large stainless steel cooking pot will do.

The cooking pan should now be placed on the electric hotplate & gently brought up to 55 – 60°C (131 - 140°F). This will cause the hot oil to bubble & spit, therefore it is important that nobody stands close enough to get burnt from the heated oil. Use a thermometer to continually check the temperature & ensure is does not burn or char by continually stirring the oil. Stirring the oil as it is heating will also reduce the chance of the oil exploding.

This is possible as if it contains any water, it will sink to the bottom (because oil floats on water) & may become heated enough to vaporise & then violently escape up through the oil, only to explode out of the top of the oil, showering the surrounding area with hot oil.

When using WVO, the required temperature is higher, but that will be dealt with the later section that deals with WVO. When the SVO has reached the required temperature, switch off the hotplate & allow the oil to slowly cool.

This is an important first step, as any water that is present in the biodiesel process will combine with the glycerol & form soap, which will be extremely difficult to remove from the biodiesel; therefore removing any water beforehand will reduce the risk of the biodiesel becoming an unusable mess. It is also important **NOT** to allow the oil to burn or char in any way, but just to heat it through because if the oil becomes overheated, it will release free fatty acids (FFA's) into the oil & it is the creation of these FFA's that must be avoided as if they are present, they will turn the biodiesel acidic, which will then damage the engine when the biodiesel is used.

This first step is known as de-watering & it should be evident that this preheating of the oil is not only an important step, but must also be achieved accurately to reduce the chance of FFA's forming in the oil which will create unwanted problems with the biodiesel.

Step 2 – Titration, titration, titration

The second step is called titration. This is the test that is performed to the oil so that the precise amount of the chemicals needed to completely turn the SVO/WVO into viable biodiesel can be accurately calculated.

To undertake the titration, the following items will be needed.

- Isopropyl alcohol (C_3H_8O)

- Phenolphthalein ($C_{20}H_{14}O_4$)
- KOH standard solution
- 1ml of the SVO that is to be processed into biodiesel
- 200ml borosilicate glass beaker
- 2x borosilicate Pasteur pipettes with graduation marks
- Borosilicate stirrer
- Pen & notepad

Pour 10ml of the isopropyl alcohol (C_3H_8O) into the 200ml borosilicate glass beaker & then add 3 drops of phenolphthalein ($C_{20}H_{14}O_4$) into the alcohol then stir it to mix.

Using one of the pipettes, start dripping, one drip at a time (0.5ml), the KOH standard solution into the alcohol. As the KOH standard solution is added, the mix will flare red, but quickly disappear as the mix is stirred.

Continue to drip the KOH standard solution into the alcohol one 0.5ml drop at a time until the red colour remains for between 20 – 30 seconds. At this point the isopropyl alcohol (C_3H_8O) has now been neutralised so this is known as the 'blank titration point', but in reality, it is the acidity of the isopropyl alcohol that has been neutralised.

Using the other pipette, add 1ml of the filtered & dewatered SVO to the isopropyl alcohol (C_3H_8O) & stir together until the oil & alcohol have completely combined together. The red colour will now have disappeared because the FFA's in the oil will have tipped the mixture from neutral to being more acidic.

From this point forward, use the pen & notepad to keep an accurate count on the amount of the KOH standard solution that will be added to the mix.

Using the original pipette (the one that was used with the KOH standard solution), drop one 0.5ml drop of the KOH standard solution into the mixture. Again, as happened previously, the mix will flare red, but quickly disappear as the mix is stirred.

Keep adding the KOH standard solution into the mixture one 0.5ml drop at a time (& stirring) until the mixture turns red for between 20 – 30 seconds. At this point, the FFA's in the oil have been neutralised.

Note down the amount of KOH standard solution it took to go from the blank titration point, to the point where the FFA's were neutralised. This should never be greater than 3ml. If it is, reject the mix, clean all the equipment & repeat the test again from the beginning.

This test should be repeated at least three times & the average of the three results should be used. Therefore, if the results were 1.5ml, 2.0ml & 1.5ml, add the three results together & divide by three.

$$1.5 + 2.0 + 1.5 = 5$$

$$\frac{5}{3} = 1.666$$

Therefore, for this example, the (average) result is 1.666ml.

As the amount of KOH standard solution that was needed to neutralise the FFA's in 1ml of SVO is now known, it is just a matter of simple multiplication to determine exactly how much will be needed to neutralise the FFA's in the remaining 1 litre of the SVO that has been heated.

The standard measurement is always 9g of potassium hydroxide (KOH) per litre of SVO. To this the average result is added. Therefore 9 + 1.666 = 10.666g of potassium hydroxide (KOH) will be needed for every litre of **this** batch of SVO.

Every batch of oil will be different, but when a KOH standard solution is used, always use 9g of potassium hydroxide (KOH) per litre of SVO in the calculation.

If the standard solution was made with sodium hydroxide (NaOH) then the titration test will also need to be made using a NaOH standard solution, then the calculation would be slightly different. 3.5g of sodium hydroxide (NaOH), not 9g of potassium hydroxide (KOH).

Therefore, using the same results, 3.5 + 1.666 = 5.166g of sodium hydroxide (NaOH) will be needed for every litre of **this** batch of SVO.

There is however advantages to using potassium hydroxide (KOH) instead of sodium hydroxide (NaOH). Potassium hydroxide (KOH) is slightly less dangerous & the chemical reaction is slightly easier. Therefore, the majority of biodiesel recipes that can be found will tend to use potassium hydroxide (KOH).

Step 3 – Methoxide magic

The next step is to make the methanol (CH_3OH) mixture. For every 1 litre of oil, 200ml of methanol will be required, therefore for this batch, just 200ml will be required. Some recipes call for 250ml per litre & it is true to say that the more that is used (up to 25% by volume of the SVO), the more complete the transesterification will be, but the more will be left to recover at the end. Therefore 200ml (20% by volume of the SVO), is a workable amount that will be effective.

Add the 200ml of the methanol (CH_3OH), then the 10.666g of potassium hydroxide (KOH) that was calculated in the titration together in a 500ml borosilicate beaker. Carefully stir both chemicals together until the potassium hydroxide has completely dissolved into the methanol. As this happens, heat will be released & the liquid will become a little warm.

The mixture that is in the beaker is now called **methoxide**.

It is a mixture of two dangerous chemicals & has not been transformed into anything safer, it is still dangerous & as you are working with it, those dangerous methanol fumes are evaporating out of the top of the beaker, so when the two chemicals have completely mixed together, carefully transfer the mix into the airtight glass container & put it in a safe place whilst the next stage is being prepared.

Step 4 – The big mix of compromises

The next step will be to mix everything into the SVO & it is imperative that this (just as every other step) is undertaken safely & precisely.

Methanol (CH_3OH) will boil at 64.7°C (148.46°F), therefore it needs to be kept well below this temperature, but the mixing process works best when the oil is at a higher temperature, therefore a compromise is needed to allow the oil to be hot enough to be viscous, but low enough to stop the methanol boiling off into the atmosphere. Therefore a temperature of around 46°C (115°F) would be a good working temperature to aim for.

Switch on the electric hotplate again & raise the temperature of the oil to approximately 46°C (115°F), remembering to stir the oil as it heats. When the oil is up to temperature add all the methoxide taking care not to splash. As the methoxide is added, the whole mix needs to be stirred. A violent stir works best but great care will need to be taken so as not to cause any splashes. Therefore there is another compromise between vigorous stirring & not splashing. The whole pot of oil & methoxide should be stirred together vigorously to ensure the methoxide fully mixes with the oil. Together they should now have a milky appearance & the whole mix should be stirred less vigorously, but continually for 30 minutes.

Some individuals use an electric drill with a stirring attachment such as a mixing paddle to undertake the stirring, but in doing so are risking an explosion as electric drills spark & the heated oil/methoxide mix will be surrounded by methanol vapours; which is an explosive atmosphere. Therefore, the only way this will be a permissible way to stir the mix is if a drill with a brushless motor is used. The more thoroughly the mixture is stirred, the better the reaction will be, so if it is in any way possible to use a drill with a brushless motor, please do so. In larger batches, achieving an effective stir will be impossible unless an electric motor is used, therefore consider purchasing a suitable drill if making biodiesel is to become a regular hobby.

Once the oil/methoxide has been stirred continually for 30 minutes, it should now have turned from a milky colour to a dark brown colour. Once this colour change has occurred, it is then time to take the mix off the heat & to stop stirring.

Once the mixture has cooled sufficiently, pour the oil/methoxide mix into a brewing bucket. It may be beneficial to use one without a tap.

Step 5 – Separation

The mix in the brewing bucket should now be left to settle overnight. Whilst it is settling, ensure that all the equipment that has been used so far is cleaned outside using the hose pipe. NEVER take anything indoors to wash in the kitchen sink.

The following day (or after a minimum of 12 hours has elapsed), it is time to examine the mixture. What should now be evident is that a separation has occurred & therefore a thick brown layer will be at the bottom of the bucket. This will be the glycerol, whilst the light coloured liquid that is floating above it will be the biodiesel.

The brown glycerol ($C_3H_8O_3$) layer at the bottom will also contain surplus methanol, the catalyst & also any soap that was formed when the FFA's were neutralised. In this 1 litre mix, there should therefore be less than 200ml of this glycerol mixture. It should be obvious by now that this glycerol mix is just the byproduct which is made up of approximately 49% methanol, roughly 49% glycerol & the remainder being made up of soap & the leftover catalyst. If there is less than 100ml of glycerol, then there is a problem. The most likely cause of this is that the reaction did not fully complete the transesterification process because not enough catalyst was used, therefore the titration results will need to be checked & the SVO reprocessed to liberate more glycerol.

If it is apparent that there is a soap layer between the biodiesel & the glycerol, then the likely cause would be either the oil was not de-watered enough, or too much catalyst was used. Using too much catalyst can also cause an emulsion or gel to form between the biodiesel & glycerol.

Providing all is well with the mix, it will be evident that the glycerol mix at the bottom of the bucket is far more viscous than the biodiesel which is floating above it, therefore providing that care is taken, it should be possible to slowly pour the biodiesel into another brewing bucket. Into one with a tap at the bottom would be best. In larger processes, the glycerol is usually tapped off from the bottom leaving the biodiesel behind.

If the glycerol is too liquid to allow the biodiesel to be poured out from the bucket, use a ladle to remove as much as possible whilst taking care not to disturb the glycerol layer at the bottom as it is important that the glycerol does not remix with the biodiesel. If it does, the mix will have to be left to settle for a further 12 hours to allow it to settle out again.

Before this 1 litre of biodiesel which has been produced can be used to fuel any motor, it must now be tested & then washed.

Step 6 – Testing

For the tests, there are some further items of equipment that will be needed.

- 200ml borosilicate glass beaker
- 200ml petrochemical diesel sample
- 200ml SVO
- 300ml tin or jug that has a 3 – 5mm hole drilled in the side wall, near to the base
- Petrochemical hydrometer
- Stopwatch or watch with a second hand
- Pen & notepad

The first test is visual. There should be a clear & distinct colour difference between the glycerol & the biodiesel. If this is the case, then you will probably have made biodiesel, but the quality is unknown at this point.

Place some of the biodiesel into a borosilicate graduated cylinder. Fill it approximately to the half way mark, then insert the petrochemical hydrometer. Top up the biodiesel to allow the hydrometer to float freely, then read off the density at the bottom line of the meniscus, not the top. The 'specific gravity' reading should be somewhere in the region of 880 & 900.

If you do not have a petrochemical hydrometer, weigh 1 litre of the biodiesel & the result should equal between 880 & 900g (but don't forget to subtract the weight of the container from the weighing scale's reading to get the true weight of the biodiesel). Return the biodiesel to the bucket.

Next, check the viscosity. Pour the 200ml petrochemical diesel into the 300ml tin with the pre-drilled hole in the side & then time how long it takes for the petrochemical diesel to flow through the hole & empty the tin. Make a note of how long it takes. Next, pour 200ml of the biodiesel into the 300ml tin with the pre-drilled hole in the side & then time how long that takes to flow through the hole & empty the tin.

Make a note of how long it took. Then, pour 200ml of SVO into the 300ml tin with the pre-drilled hole in the side & then time how long that took to flow through the hole & empty the tin. Make a note of how long that took. The results should show that the biodiesel emptied from the tin quicker than the SVO did, which will indicate that the viscosity of the oil has been reduced & the biodiesel time will be similar to (but take little longer) than the petrochemical diesel, showing that the biodiesel is ever so slightly more viscous than the petrochemical diesel.

If the biodiesel is too viscous, the fuel pump in the engine will have to work harder & therefore it will break or wear out far sooner. If the biodiesel is very viscous, the fuel pump may not pump at all, therefore, the correct viscosity is an important quality of biodiesel. It should be similar to the petrochemical diesel.

If all is well with the biodiesel, the next stage is washing.

Step 7 – Washing the biodiesel

For this, the following equipment will be needed.

- At least two brewing buckets. One with a tap
- Access to water
- 500ml measuring jug or borosilicate glass beaker
- Litmus paper
- Ladle (optional)

Despite the fact that the majority of the unwanted by-products are in the glycerol, there is still a lot floating around in the biodiesel, contaminating it. Approximately 3% of the biodiesel is still methanol (CH_3OH) & as methanol is a solvent it has captured soap & everything else that is left in the mix. Therefore, if it were used in an engine in its current state, it would contaminate the fuel lines & the engine, which will obviously lead to problems.

Therefore the biodiesel will now need to be **washed** to remove these unwanted contaminants. This is achieved with water (H_2O) as water will absorb the methanol (CH_3OH) & therefore take all the impurities with it, but it will take several washes before the biodiesel will be safe enough to use.

Tilt the bucket onto its edge & as gently as you can manage, pour 300ml of water down the inside of the bucket. What should happen if you have been gentle enough is that the water has sunk to the bottom of the bucket. This happens because the biodiesel is a little less dense than the water.

- Specific gravity of biodiesel 880 – 900
- Specific gravity of water 1000

This is the reason that oil floats on water. Very gently stir the biodiesel & water in the bucket. This should not be vigorous stirring, but just enough so that the water layer & the biodiesel layer mixes together. If the stirring is too vigorous, the soap will bind with the biodiesel & the result will look like chicken soup. Turning biodiesel into something that looks like chicken soup should be avoided at all costs.

What should happen is that the water has become cloudy. This is happening because the methanol & all those impurities that it's holding on to are now dissolving into the water.

Once it has settled, the water can be removed. If you are using a bucket with a tap at the bottom, you can use the tap to remove the water from the bucket. If the bucket has no tap, use a ladle to gently remove the biodiesel & put it into another bucket. Now it's time to add another 300ml of water. As previously, tilt the bucket & dribble or pour it very gently down the inside of the bucket. Again, as previously, the water will sink to the bottom & then it can be stirred again. This stirring can be a little more energetic this time as there will be far less soap contaminating the biodiesel. Once it has settled, the water can be removed again & the whole process repeated in total approximately seven times.

On each refill of water, the mix can be stirred with more vigour as there will be less & less soap contaminating the biodiesel. As this happens, the ability for biodiesel to absorb water will also lessen. Therefore, the water is not only cleaning the biodiesel, but it is also drying it too. Another thing that the water washing achieves is that it is removing the methanol at the same time; therefore, that methanol can no longer contaminate the biodiesel with any more glycerol or soap. So when it's removed from the biodiesel, it's removed for good.

If you find that the biodiesel has been turned from a transparent golden colour into a cloudy yellowish liquid, it is because there is now water mixed in with the biodiesel. If it is left for a short while, the water molecules will sink to the bottom & the transparent golden colour will return. Just remember to remove the water afterwards.

Next, test the pH level of the biodiesel by dipping a strip of litmus paper into it, or by using an electronic pH meter. Ideally, the biodiesel should have a neutral pH, which is pH 7. If it is much higher, then the biodiesel is too acidic, therefore repeat the washing process a few more times to reduce the pH.

Step 8 – Filtering & de-watering

During the process of making biodiesel, it is possible that it may have picked up some contaminates & even though the diesel engine will have a fuel filter, it will not be advisable to use the biodiesel to fuel an engine until it has been filtered.

Some individuals pour the biodiesel through a j-cloth that is stretched over another bucket, but the following method will achieve far better results. For this, you will need the following:

- Automotive in-line fuel filter
- 2 metres of biodiesel compatible fuel line hose
- Small funnel
- Storage container for biodiesel

166

Automotive in-line fuel filters can be bought online very cheaply. It is possible to get a 10 pack for less than £20 or just one for about £3. It will have an opening at each end. One is the inlet, one is the outlet. If the filter has 6mm fittings, then use with a 6mm biodiesel compatible fuel line hose, if it is 8mm, then use with an 8mm biodiesel compatible fuel line hose. Cut the 2m hose into two & attach a length to each end of the filter. You should now have a 2m length of hose with an in-line fuel filter in the middle. Finally, fit a small funnel to the open end of the hose that is attached to the inlet side of the filter.

The unattached end of the pipe can now be placed inside a suitable fuel container & the biodiesel can be poured into the funnel. In-line fuel filters are fine for gravity fed fuel systems, therefore the biodiesel will flow, or drip through & the filter will remove all the contaminants that can block a fuel filter fitted to a car engine. The resultant biodiesel should therefore be clean enough to use to power an engine.

If the in-line filter becomes blocked, it is a simple matter of removing it & replacing it with another.

At this point, if the biodiesel looks in any way cloudy, then it may have absorbed water from the atmosphere. The range of permissible water content for biodiesel is between 1,200 – 1,500 PPM (parts per million). Anything above this threshold will cause engine damage, therefore if after the wash step the biodiesel remains cloudy for a day, then it is definitely contaminated with water & therefore needs to be de-watered. To do this, the following will be needed.

- Electric hotplate
- Stainless steel cooking pot (or maslin pan)
- Thermometer

Pour the biodiesel into the maslin pan & place on the hotplate. Heat the biodiesel to gently bring it up to 55 – 60°C (131 – 140°F).

This will cause the hot biodiesel to bubble & spit, therefore it is important that nobody stands close enough to get burnt from the heated biodiesel. Use a thermometer to continually check the temperature & ensure is does not burn or char by stirring the biodiesel. Stirring the biodiesel as it is heating will also reduce the chance of it exploding.

This is possible as if it contains any water, that water will sink to the bottom (because oil floats on water) & may become heated enough to vaporise & then violently escape up through the oil, only to explode out of the top, showering the surrounding area with hot biodiesel. After 20 minutes of this heating, the biodiesel should be left to cool. After an hour of cooling, it can be filtered & stored for use.

Step 9 – The clean up operation

With this small experiment in biodiesel production, the by-products that you will have left over are only small quantities, therefore they are relatively easy to deal with, but if the production is scaled up to any degree, dealing with the by-products are completely different, therefore this section only deals with these small quantities. Larger amounts of the by-products will be dealt with later in the book.

For this, the following equipment will be needed:

- two flexi tubs

The glycerol that was left over after step 4 will contain soap, methanol & contaminants. This should therefore be transferred to a flexi tub & left outdoors.

As domestic soaps contain glycerol & contaminants, this can be set aside for soap production which will be detailed in a later chapter of this book.

The water that remains from the wash process contains methanol, soap & small quantities of all the chemicals that were used, therefore the water should first be left outdoors for several days in the fresh air. This will give the methanol time to evaporate slightly, then afterwards, the remaining water can be flushed down the drain as it is only a small quantity & the methanol will have had time to completely evaporate.

Now clean all the equipment that has been used since the last clean up, but do not do it in the kitchen sink, clean everything outdoors.

Test batch №2 – WVO

This is the second batch & can again be thought of as another test batch, or practice batch. It will create another 1 litre of biodiesel, but this time it will be made from WVO. It differs very little from the process' that was encountered in test batch №1, but there are differences & they are important differences.

The first & most obvious difference is the oil, as it is possible that the WVO that has been selected may have been sitting around for a while. This is a good thing as the heavier fats will have had time to sink to the bottom. Also, even if the WVO looks like it doesn't contain any solids, it will have some, therefore the WVO should now be passed through several sieves to remove these unwanted particles. Any sieve will do, but the finer they are, the better the final result will be & also the slower the sieving will be. Oil may move through a large sieve rather quickly, but only very slowly through a small grade sieve. If the sieve is only microns in size, it may take several days for the oil to work its way through.

One Heath Robinson method of sieving that many adopt is by having two 25 litre home brew buckets with one having a j-cloth stretched over the top. Oil is then poured from the open bucket onto the j-cloth covered bucket.

The j-cloth acts as a sieve, removing the larger particles. This may be an attractive method until the scale of production & proficiency of the biodiesel maker calls for more suitable sieves. In that instance, laboratory sieves will be more useful, far slower, but far more effective. Also, it may be possible to construct a DIY sieve for anyone with suitable skills as the mesh used in these sieves are available to purchase at a greatly reduced price. It may even be possible to incorporate several sieves in a tank to tank system, where the WVO is pored into one tank, where it is then fed through several sieves, then finally ends up in another tank where it is ready to use. Which method is used is unimportant. What is important is that the WVO should be sieved effectively before it is used.

The equipment

The same items that were used in test batch №1 will again be needed for this second batch. Also, the KOH standard solution that was made for the first batch will also be used here.

Step 1 – Safety first & de-watering

Step 1 is exactly the same as it was in test batch №1, except after the oil has been sieved, it must be heated to 100°C (212°F) for at least 20 minutes before being allowed to cool.

Step 2 – Titration, titration, titration

Step 2 is exactly the same as it was in test batch №1. There will be a difference in the titration results as the WVO will contain more FFA's & these will alter the results, therefore the amount of the KOH standard solution that will be needed to neutralise the FFA's will be greater than was seen in test batch №1.

As previously, it will be 9g of potassium hydroxide (KOH) per litre of WVO, & the average to be added from the test result.

Step 3 – Methoxide magic

The methoxide step is exactly the same as it was for the SVO.

Step 4 – The big mix of compromises

Step 4 is exactly the same as it was for SVO.

Step 5 – Separation

The separation step is exactly the same as it was for the SVO.

Step 6 – Testing

Step 6, the test stage, is exactly as it was for the SVO. The only difference is that when the viscosity test is undertaken, ensure that a WVO sample is used, not the SVO sample.

Step 7 – Washing the biodiesel

Washing the biodiesel is exactly the same as it was for the SVO.

Step 8 – Filtering & de-watering

Step 8, filtering & de-watering will be exactly the same as it was for SVO.

Step 9 – The clean up operation

Step 9 is the same as it was for SVO.

Conclusion

Hopefully, if every step of the SVO &/or WVO production process was followed to the letter, 2 litres of biodiesel will have been created using this tried & tested method.

As well as creating some useable biodiesel & helping to save the planet, an appreciation of organic chemistry may also have been cooked up. There may even be questions that have been formulated too, such as what went wrong with the batch, what is the best way to store the biodiesel, how to make a big batch or how to make soap form all that glycerol? All these questions will now be answered in the coming chapters.

Chapter 10 – Scaling up production

After making one or two test batches, the obvious thought is to now ramp things up & start making bigger batches. After all, it is just a case of scaling everything up, but before this is tackled, there are some points that need to be examined.

Longevity of the biodiesel

Just as a loaf of bread has a shelf life, so too does biodiesel. Biodiesel will only last for up to 5 months. There could be a noticeable drop in quality after 3 months. Therefore, it is not advisable to keep any more than 3 months supply of biodiesel at any one time.

The longevity of the biodiesel & how to ensure it is in optimal condition is covered in more detail in chapter 13, which covers the storage of biodiesel.

A taxing conundrum

If you are ever in a position where you have too much biodiesel to use & need to slim down your stock before it goes off, the following should be considered.

If biodiesel production in the UK is undertaken on a scale where it can be classed as 'domestic & for own use', then legally no laws are being broken, but if the scale of production increases to a level where it is necessary to sell it, or even give it away, then it can no longer be classed as being domestic or for own use. Therefore the long arm of the law will come into play.

Firstly, a permit to operate should be obtained from the Environment Agency. The application is currently free under Exemption U5 – using biodiesel produced from waste as fuel, but must follow an assessment which currently costs £750 (+vat) which is a paperwork exercise where you 'justify & demonstrate' that you are turning a waste product into a useful, saleable product. You may need to also apply for an environmental permit as your operations could pollute air, water or land. There are also a number of laws that need to be adhered to, such as *The Control of Pollution (Oil Storage) (England) Regulations 2001*, but providing production is less than 2,500 litres of biodiesel per annum, accurate records are kept & that you comply with any requirements from the Environment Agency, then the biodiesel produced should be exempt until you produce more than the 2,500 litre threshold. Until that point, you are classed as an **exempt producer**.

You may however be required to provide samples to prove that the biodiesel complies with 'their' legal definition of what biodiesel currently is. That is it must comply with EN14214 & the main points are (see chapter 17 for the full list):

- The sulphur content does not exceed 0.005% by weight.
- The ester content is not less than 96.5% by weight.

If you go over the 2,500 litre threshold, there will be duty to pay on any biodiesel you make & use as fuel for a road going vehicle. Yes, even your own vehicle, even if it is used to drive on your own land (such as a tractor in your field), it is still taxed, but at a lower rate. If it is driven on the public highway (any road or bridleway in the care of the local authority) then the higher fuel duty should be paid.

The current excise duty for biodiesel used on a road is[51] £0.5795 per litre, which is exactly the same as petrochemical diesel & petrochemical unleaded petrol.

---

[51] https://www.gov.uk/government/publications/rates-and-allowances-excise-duty-hydrocarbon-oils/excise-duty-hydrocarbon-oils-rates - 24/07/2020

If it is used solely off road, then it is taxable at £0.1070 per litre, which is the same rate as any biodiesel used for heating oil is currently set at. Therefore, assuming you have reached your 2,500 litre threshold & you then made 1 litre of biodiesel from SVO & the 1 litre from WVO, then you will owe HMRC £1.16p in tax if you plan to use the biodiesel in your car, or 21½p if you intend to use it to heat your home.

HMRC accepts payments as follows. To pay by Bacs (Bankers Automated Clearing System) or CHAPS (Clearing House Automated Payment System), use the following bank account details:

Account name:	HMRC General BT Receipts acc.
Sort Code:	20-05-17
Account №:	40204374
BIC:	BARCGB22
IBAN:	GB86BARC20051740204374

Cheques can be made payable to 'HM Revenue & Customs only' sent to:

HMRC Direct
BX5 5BD

If the threshold of 2,500 litres has been met, you are no longer an exempt producer, so must now register with HM Revenue & Customs using form **EX103**[52] (this is for private individuals, use **EX103A** if a business). All records must be kept for a minimum of six years & submissions must be made monthly using form **HO930**, but you must register first. The government has issued guidance on the tax situation called *Excise Notice 179e: biofuels and other fuel substitutes*[53].

---

[52] https://public-online.hmrc.gov.uk/lc/content/xfaforms/profiles/forms.html?contentRoot=repository:///Applications/Excise/1.0/EX103&template=EX103.xdp - 24/07/2020

[53] https://www.gov.uk/government/publications/excise-notice-179e-biofuels-and-other-fuel-substitutes/excise-notice-179e-biofuels-and-other-fuel-substitutes - 24/07/2020

You need to read & comply with the requirements of 179e & be aware that the whole tax situation could change at a moments notice & therefore if any doubt exists about the legality of the biodiesel produced, the quantities that are permissible & the duty that should be legally paid, then at the very least check online with other small scale/hobby biodiesel producers or seek the services of a solicitor who specialises in excise duty.

It is down to the individual to pay the correct rate of duty & there are big fines, penalties & potential prison sentences waiting for those who do not pay up & then get caught.

It would also make sense to get vat registered, just so that the vat can be reclaimed from the tax office.

As well as tax, in the UK, there are also numerous other legal aspects to contend with. The first of which will be local government.

Planning permission

If the decision is made to step up from domestic production to a commercial sized operation, your local authority will insist on a change of use from the existing domestic categorisation to an industrial categorisation.

A change of use application will therefore need to be made from class C3 (dwelling house), to either B1(c) (business light industrial), or B2 (general industrial). There is no planning authority in the UK that would entertain this on a domestic premises, therefore suitable industrial premises would need to be sought.

Environment Agency

Taking a step up to commercial production will mean that the **exempt producer** moniker can no longer be used, therefore the full force of the Environment Agencies remits will come into force.

The first step will be to obtain from the EA an Integrated Pollution Prevention & Control Licence (IPPC). The permit application is currently set at £3,032, it has a minor variation fee of £909 (for an amendment of a minor technical nature) & a normal variation fee of £1,516 (for any amendment larger than minor). An application to transfer the license costs £2,459 & to surrender it costs £1,819. Consultants may also be needed to ensure that the technical paperwork is filled out correctly the first time, or a few minor amendments will soon add up.

Collecting waste oil also entails another licence. A Waste Carriers Licence. These cost £154 for three years, then £105 for the next three years. Any organisation which fails to register is liable to a fine of £5,000. After obtaining this license, the WVO can only be taken to a facility with an IPPC licence. If you have been granted an IPPC licence, then this will include your little biodiesel facility.

The Health & Safety Executive (HSE)

Ramping up production will also involve the HSE. They would not involve themselves in small scale domestic operations, but stepping up from small scale will now entail an involvement to some degree with the HSE.

In the UK, the HSE produce guidelines for safe working practices. Some of these can be obtained for free from their website, but most of the guides are obtainable at a cost. It may therefore be worth checking with your local reference library as they should have a subscription & therefore allow you to access the material for free.

The entire biodiesel production facility will need to be checked by a competent person (an engineer with experience & adequate liability insurance).

There is no longer a requirement to register the facility, but if construction involves more than 500 person days or involves 20 workers for more than 30 days, then the HSE must be informed using form F10[54]. There are no longer any other notification requirements beyond that, although the HSE can turn up at any time, unannounced, to check that ALL their requirements are being complied with.

Insurance

Anyone operating at this level will need insurance. Insurance for the biodiesel production facility (even if it is two oil drums), insurance for the premises, personal liability & the liability of the biodiesel itself. If anyone is employed, then employers liability insurance will also be needed. This could all cost upwards of £50,000 per annum.

Trading Standards

If biodiesel is being sold, then if any pump is used to sell it (such as a forecourt type set up), then the pump must be calibrated by Trading Standards (at an hourly cost) & of a type approved by Weights & Measures. The cost of these pumps can start at approximately £5,000.

The pump must then be tested & calibrated to the Measuring Instruments Directive 2014/32/EU & the *Measuring Instruments Regulations 2016*.

---

[54] https://www.hse.gov.uk/forms/notification/f10.htm - 27/07/2020

Any error in selling liquid fuel can not equate to more than 2p by law, therefore it will be far safer to 'sell' set amounts in your own containers, such as a 20 litre drum, or a 205 litre drum. Heavy penalties exist for those who sell short measures.

A big wash

If all the legal issues have not put you off from going beyond the 2,500 litre barrier, then there are other things to consider, such as washing the biodiesel. So far, only two, 1 litre mixes have been undertaken. If the mix uses 5 litres at a time, the same wash that was undertaken in step 7 would still be permissible. But when the mix goes up to 10 litres & beyond, moving large heavy buckets will start to become more difficult. At 20 litres it will start to get very dangerous & beyond that, it will start to become impossible.

Therefore different washes will need to be undertaken as it will be impossible to tilt large buckets or heavy oil drums. There are therefore several options available for large scale washes.

Static water washing

Static or gravity washing is the most simple of all the water washing techniques. It involves pouring water into the same tank or drum as the biodiesel, the methanol & its impurities will then migrate into the water over a number of hours.

The whole process takes place at the boundary between the biodiesel & the water without any need for any physical interaction.

This wash is usually undertaken over a 24 hour period & tends to be used as a gentle first wash to remove the soap that can cause problems during vigorous washing process. It is a highly effective washing method for removing the excess soaps that are present when oils with high FFA's are used.

Mist washing

Mist, or spray washing as it is known is the process of spraying a mist of water over a tank or drum of biodiesel. The fine water droplets will then drop down through the biodiesel, collecting the methanol & its impurities as it sinks to the bottom. The benefit to this method is that it will collect any methanol vapours that have collected above the biodiesel, but it is more aggressive than the static wash, so should not be used before the majority of the soap molecules have first been removed.

There is also a risk of the tank overflowing unless either the water is recirculated, or a cut-off device is installed. Also, if the water is not recirculated, there is the potential to use a high volume of water.

The size of the water droplet will also have an effect on the mist washing process. Large water droplets will agitate the biodiesel enough that if there is any soap present in the biodiesel, then it will create an emulsion. Raising the spray head to a height of at least 300mm above the surface of the biodiesel should be enough to reduce this risk. Small water droplets will also cool the biodiesel enough to affect the process; therefore a heat source is best used to mitigate this effect. If the tank is sealed, this cooling problem will resolve itself as it takes place in a sealed tank when the air above the biodiesel becomes saturated with water vapour. In the summer months, feeding the spray head through a black hose pipe laid out in the sun may raise the temperature of the spray enough to reduce the heat loss problem. A happy medium between fine & large droplets would be rain drop sized droplets.

Pump washing

Pump washing is a very harsh system of washing which will only work if the biodiesel is well made & well processed.

Biodiesel that contains under processed oil containing diglyceride or monoglyceride molecules will turn the mix to an emulsion. If the biodiesel is well processed, then this wash will split the biodiesel & water into very fine particles, which creates a greater surface area & therefore cause a better interaction between the water & the biodiesel. As this is a very aggressive wash process, it should therefore never be attempted until at least one static wash has been completed & also at least three mist wash processes have also been completed. Pipe work fed from the bottom of the tank & discharging at the top of the tank must also be constructed beforehand, along with the inclusion of an electric pump to pump the water/biodiesel mix through the pipe.

In pump washing, the water from the last wash (the last mist wash) is left in the tank & it is used for the first pump wash. This is now pumped from the bottom of the tank along with any biodiesel to the top of the tank & allowed to fall harshly onto the top surface of the biodiesel. It is best practice to add 5% warm water during this process. If 200 litres are being washed, then 5% will be 10 litres. Run this cycle for 15 seconds, then allow the tank to rest & settle for 1 hour. The 15 seconds is just a guide. If an underpowered pump is used, then maybe 30 seconds will be needed, if a more powerful pump is used, then perhaps 10 seconds will be sufficient. Therefore, all the times detailed here are only a guide. It will take an amount of trial & error before you are able to ascertain the optimal times for the pump & system you have designed for yourself.

After the tank has settled for an hour, drain a minimum of 80% of the water that was used, & then replace with fresh water & another 5% warm water, then run the pump again for 30 seconds. Switch off the pump & allow the tank to rest & settle for a further 1 hour.

Following this, remove a minimum of 80% of the water that was used for the wash, add 5% warm water as before (but do not replace the 80% water), then run the pump for 15 minutes. Switch off the pump & allow the tank to rest & settle for a further 1 hour before removing **all** the water.

Remove 100ml of the biodiesel & place in a 500ml kilner type jar along with 100ml of warm water. Close the lid & shake up the jar vigorously to mix. After the water & biodiesel has separated, remove the water & check its pH level. Water should be pH neutral, which is 7 pH. This is also what the water removed from the kilner type jar should read if all the soap & other impurities have been removed from the biodiesel. If the water sample is over 7 pH, then further washing is required, therefore repeat the last wash, but this time add the same amount of water that was removed after the last wash, along with 5% warm water before you run the pump for 30 minutes (never pump for longer than 30 minutes) before switching the pump off & allowing 1 hour for the mix to rest & settle. Following this, remove **all** the water & repeat the 100ml shake test & the pH test.

Bubble washing

Another wash type is called bubble washing & is another aggressive wash cycle. As such, it can only be used following static & mist washes. In this wash, water is added to the tank & bubbles are then created in the water layer. These bubbles form in the water; therefore as they travel upwards through the biodiesel, they will be coated with water. The bubbles then burst at the surface & the water will sink back down through the biodiesel. The water therefore interacts with the biodiesel as the bubble travels upwards, then again as it sinks back down. Small bubbles with a low flow rate are not as aggressive as larger bubbles with a high flow rate.

This should only be used in a sealed tank as a high level of methanol vapours can be liberated from the biodiesel during the wash, but it does use far less water than other techniques.

Any bubble stone that is designed for use in an aquarium will dissolve when it comes in contact with biodiesel, therefore it will be better to remove the bubble component & just keep the pump.

Use the type that allows you to control & adjust the air flow, then attach this to a clean wet stone by drilling a hole into it so that a polyethylene tube can be inserted. Attach the other end of the tube to the pump. The stone may need to be weighted down to stop it from floating up through the tank. If you can not find a suitable wet stone, block the end of the polyethylene tube, & then create lots of holes in the last 150mm with a pin, then weigh down the end of the tube at the bottom of the tank.

The mix should be of 33% water, 66% biodiesel. Gently bubble for between 6 – 8 hours & then allow it to settle for 1 hour. Following this, remove the water. Replace the water & again bubble for between 6 – 8 hours before allowing the mix to settle for 1 hour before removing the 2nd lot of water. Repeat the whole process for a 3rd time, but when the water is removed for the 3rd time, undertake the same pH test that was detailed at the end of the pump washing process. If the pH result is 7, then all is well, if not, repeating the bubble wash process for a 4th time.

All these washing techniques use water, therefore the biodiesel will therefore have an amount of water mixed into it & so it may look a little like orange juice; therefore all this latent water must first be removed before the biodiesel can be used in an engine. It is removed by drying the biodiesel.

Drying biodiesel

Removing the residual water left in the biodiesel can be achieved by simply allowing the biodiesel to settle over a few days because the water will always settle to the bottom & the biodiesel will always float on top. This is the simplest method & is called 'settling'.

If this is undertaken in a dry climate, then all is well, if you do not live in a dry climate, then it may not always be 100% effective because if the batch contains any under processed oil which contains diglyceride or monoglyceride molecules, these molecules will literally suck moisture from the air. It is because these molecules are 'hygroscopic'. If the biodiesel is poorly washed, it will also contain soap &/or methanol & these will cause the same hygroscopic effect to occur. The more humid the atmosphere, the quicker this will happen. It is even possible that the biodiesel may be sucking moisture from the air at the same rate that water is settling out, or even, absorbing more than is settling out, but this will only happen in very humid conditions.

If this hygroscopic effect does occur, it is still possible to dry the biodiesel, but the drying process must become more vigorous so that the drying outweighs any hygroscopic effect.

This can be achieved by using a combination of actions that can be called bubble drying. Here, heat the oil to a temperature of approximately 55°C (131°F). When the biodiesel has reached this temperature, insert the bubble wash apparatus & switch it on so that the water coated bubbles will once again travel to the top of the barrel or tank. The raised temperature of the biodiesel & water will now help the water evaporate as it reaches the top, rather than dropping back down through the biodiesel as it did during the bubble wash process.

However, as there will be certain distance between the top of the biodiesel & the top of the barrel or tank, the water vapour that has escaped may therefore just sit above the surface & then be reabsorbed into the biodiesel. Therefore, an ingenious solution can be made by making a lid for the barrel or tank that contains two holes approximately 100mm in diameter. At one hole fit a fan. An ideal sized fan would be of the type used in a home computer as this is not too aggressive, but gives a steady stream of air. Another alternative is one that is designed to pump moist air from a shower room.

Switch on the fan so that it blows clean air into the moisture laden air that is in the air gap above the biodiesel & then the moisture laden air will be pushed out through the other hole.

This process could take 10 minutes, or several hours depending on the water content & the environmental conditions. Therefore, at intervals, remove a sample (100ml should be fine) & place in a glass container (use a laboratory beaker) & the sample should be slightly golden & clear without a head on top, or be floating above a water layer. It should also be possible to read a book, or newspaper through the biodiesel without any effort.

There are numerous other methods used to dry the biodiesel, but the above method should work the best. If an alternative method is needed, there are numerous internet sites that offer many alternative methods. It will be a case of finding the best method to suit your unique setup.

Scale of setup

It would be best to decide beforehand what the intended scale of the production should be, as it will allow equipment to be sourced that will fit with the intended production. It will be no use dealing with 5 litre buckets when a 199 litre oil drum would be more suitable. Therefore, it will be worth looking at various setups to see what can be constructed, or purchased.

Self build, own design

When designing a unique system to your own exacting specifications, there are many factors to consider. When scaling up from the small 1 litre tests, most individuals create their own set up which they construct from 205 litre oil drums.

There is nothing wrong with this setup, but consideration must be given to the 2,500 litre limit & also the useable three month biodiesel lifespan. If 200 litres were to be created, would it be used within the three months? If not, will it be possible to brew smaller batches in the same equipment?

It is possible to brew smaller batches using these oil drums, although to do so, it should be obvious that there will be some limitations. For instance, if the biodiesel were to be dried using the bubble method detailed earlier in this chapter, then the excess methanol & water vapours will need to be driven off with a fan as was described, or the vapours will just sit in the top of the drum being unable to escape. If 25 litre batches would be required, then smaller drums should be obtained.

It may not be possible to use a selection of drums in different sizes as to heat the oil, an electrical heating element will need to be fitted within the drum, along with openings for pipe work. Therefore, a suitable size should be selected that will allow for the creation of the right sized mix for your exact requirements.

UN approved steel drums are available in 25, 110 & 205 litre sizes & cost around £30 for the 25 litre size, £80 for the 110 litre & £95 for the 205 litre size.

The benefit to using these standard sizes is that they are usually constructed from either carbon steel or stainless steel & are therefore relatively easy to cut & weld. There are barrels made from HDPE, but these are not as easy to modify & also not UN approved, but are excellent storage tanks & are therefore used in the chemical & food industries.

If a competent welder is found, it should be a simple job for him/her to weld a convex base & a stand to the bottom of an oil drum, but a simpler & cheaper method may be just to hammer a convex shape into the bottom. NEVER use an old central heating boiler as it will be made from copper & it will react to the biodiesel. If it is only possible to get carbon steel or mild steel drums, then the inside should be coated with chemical resistant paint or a liner before being used.

HDPE tanks with conical bottoms are also available. These have the benefit of being pre formed, have a graduated scale on the side, they are easy to cut & semi transparent so the contents can be easily examined, but will not be suitable for the inclusion of an immersion heater.

Pipe work should be constructed from a fluroelastomer material such as Viton. Lengths of these can be obtained from the internet & it is sold by the meter. Compression fittings should be used that are not made from copper. Again a fluroelastomer material would work best & should be sealed with silicon sealant or PTFE tape.

A multifunction pump will be needed to move liquids from tank to tank. An enclosed pump should be used that has a bronze impeller. When completely plumbed in, a system of taps will allow liquids to be simply & swiftly transferred from tank to tank.

The heating to one of the tanks should be undertaken with a domestic immersion heater that can be fitted low down in an oil drum or tank after a suitably sized hole has been drilled.

Any large tank that is heated will also need to be insulated. The surface area of a tank or drum is rather large & therefore it will lose a lot of heat.

The inclusion of ventilation must also be incorporated into the tanks, or the methanol will be trapped & create an explosive & toxic atmosphere. This should always be vented to the outside.

All the pumps will need to be constantly primed, or they will not pump & the mixer will need to be efficient for the tank that is used. The wash tank must also allow for access for bubbling & there are numerous other factors that will need consideration when designing a biodiesel facility.

Therefore it can be a difficult task, starting from scratch & designing a system that will allow you to achieve quality results, time after time, but it is not impossible. Many individuals have constructed perfectly adequate processors by themselves & on their first attempt.

Benefits will include the kudos of knowing that you have designed & constructed the processor yourself. You will also have a complete knowledge of the system & therefore understand exactly how it works, so you will know instantly if an element is underperforming & also know what will need to be upgraded or which element needs modification.

But designing & building your own system can take many hours studying designs before you are able to design your own. Parts will then need to be sourced. All of this can take many months & then there are the months that are needed to fine tune the system, with no support or advice. Do you have the necessary skills to do this? Any mistakes that are made here will definitely cost money & as volatile chemicals are used, these mistakes could cause a fire or an explosion.

The alternative would therefore be to buy a tried & tested design from the internet.

Self build, others design

There are many individuals & companies who will supply a complete system in kit form. Here, the system will be a tried & tested system that will be supplied, unbuilt. It is just a case of following the instructions & building the processor. Perhaps you may find just the instructions being sold, & therefore will need to source your own materials. The obvious advantage over the previous example is that someone else has completed the design, but whether it is a good design will be unknown.

But the advantage is that if you are new to biodiesel manufacture, there will be elements that you are unsure about, or may forget. This option should therefore cover everything, so theoretically, once it is constructed, it should work with minimal adjustments & save a load of time. The whole design should also be much safer than a self design by a newbie.

Unfortunately, the instructions may not be brilliantly clear & any ambiguity could lead to errors in the final build. If you were to need to buy your own materials, then problems could occur if you were unable to source the 'exact' materials that were specified in the instructions. For instance, if smaller hoses were the only suitable hoses available & therefore used, the pumps could be put under stress & could therefore fail.

Technical support may be limited, so you would be stuck, not knowing how to resolve the issue & not have a working knowledge of the system as it is not your design.

Some of these kits or instruction packs can also be poorly designed to start with, but how would you know beforehand?

There is therefore a risk opting for one of these kits. It would be far better to study this book & a few others to design & build a system, rather than blindly rely on a poor system & not know enough to resolve any problem issues. However, if this is the route you do take, then research the company beforehand & try to speak to individuals who have constructed the processor before you. They are the only ones who will be in a position to tell you if it is worth doing & also be able to offer advice that will be specific to your build. Therefore ask around in internet biodiesel talk rooms as much as possible.

If DIY is not a strong point, there is another solution.

Off the shelf, turn key system

Here, it is possible to purchase a complete system that contains every element needed to make a biodiesel processor, pre-assembled (or with minimal assembly) & ready to go. It will just need installation in a suitable location. This convenience does however come at a sizeable cost if bought from new, but second hand systems can be bought online. This type of system will obviously save time, so theoretically it should be possible to start biodiesel production, the same day that the processor is installed. This may be a good option if design & DIY are not an option. This type of system tends to look more professional, but that is no guarantee of success.

Checklist

As a basic check, any kit or complete installation up should contain the following as a minimum. Therefore it will be worth checking on these items:

Hardstand

Your installation may be heavy. When filled with oil & biodiesel it will weigh far, far more, therefore will a reinforced concrete hardstand be needed to support the processor. If so, this will need to be constructed beforehand & given time to cure before anything can be installed on it.

Tanks

Only purchase tanks that allow for it to be completely or partly drained. A flat bottomed tank will not facilitate this; therefore only tanks with conical bases should be used. Do the tanks allow for venting? Any tanks used for mixing the methoxide, processing or dewatering must be vented to the outdoors because of the dangers caused by the vapours. Are the tanks steel or HDPE? Perhaps it may be possible to obtain stainless steel tanks which would be the ultimate tank material.

Parts

It will be worth checking that a kit, or turn key system is supplied with EVERY component. If not, is it possible to buy those parts & will any alternative parts be compatible with everything else?

Insulation

Have the tanks & associated pipe work been given adequate insulation to reduce the heat loss, or will this need to be installed or even upgraded?

Pipes & hoses

Are the pipes & hoses made from suitable materials? Are they nylon, Viton or made from another fluroelastomer & suitable for biodiesel use? Most home biodiesel set ups use PVC hoses as it is possible to see liquids flowing through them, but they must be thick hoses or they will degrade somewhat quickly. Also, it will be advantageous to use reinforced hoses as a rupture could potentially shower the surrounding area in hot oil, methanol or methoxide. Any valves or fittings that are to be attached should be brass or iron. Never use anything that is galvanised.

Filters

Filtering is important, as was detailed earlier, therefore does the design allow for filtering the ingredients before inclusion into the mix & again at the end? Consider adding in-line filters after the dewatering tank at the very least.

Heating

How will the oil be heated? Are there sufficient heating elements to dewater the oil in the first tank, in the processing tank & also the washing tank? Heating the oil at every step is the key to making quality biodiesel. Certain heating elements are far too harsh & can burn the oil or even cause fires. It is therefore best to make enquiries in online biodiesel forums to check on which heating elements work best with which particular tank that has been adopted.

Thermostats will also need to be designed into the heating system so that the heating element is switched on at a certain low temperature & off when the temperature reaches a given level.

Thermometers also need to be included to check that the required temperatures are attained inside the tanks & that heat loss is not a problem. Also, the inclusion of timers would also be beneficial, so that the mix is kept at the correct temperature only for as long as is necessary. Never heat oil above 55°C (131°F) in the processing tank, or above 64.7°C (149°F) after the methoxide has been added as that is the temperature that the methanol boils, therefore it needs to be kept well below this level.

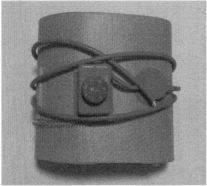 If oil drums are used, there are oil drum band heaters available that wrap around the outside of the barrel. It would be best to obtain one with a timer & thermostat. These are made from a flexible silicone belt, therefore the correct length should be bought for the oil drum it is to be used with.

Figure 40 Flexible oil drum heater (Public Domain © 2020)

It may be beneficial to use two or three of these belts under any insulation as they can be slow to heat a tank.

Pumps

Pumping the biodiesel mixture is a complicated subject. An under powered pump will work pumping between tank to tank, but if it is used to mix a 200 litre tank, then it needs to be of a sufficient size & power. Glycerol is far more viscous than biodiesel or oil, therefore the pump must be able to cope with this extra load that it imposes on the system.

A pump able to operate at 2,500 litres per hour should be purchased, or the pump will not be powerful enough to operate effectively. All pumps will need to be primed before use. If priming the pumps before use is not an option, then will it stand up to the rigours of having biodiesel or methanol sitting in it until it is used next & will the presence of methanol fumes cause a fire hazard? If the motor uses brushes it will spark, therefore it should ideally be a 'totally enclosed fan cooled' TEFC pump.

The position of the pump is also important as if it does need to be primed, then access will be needed.

Methoxide

Mixing the methoxide in large quantities presents problems as the mix of the ingredients must be complete before it is introduced into the processing tank. Therefore is there a small mixing tank included in the design to undertake this task? If not will the methoxide be mixed in a glass demijohn (aka carboy) then poured into the processing tank? If a small mixer is included in the design, how will it be possible to check on the completeness of the mixture before it is moved to the processing tank? Also, if it was mixed manually, it would be dangerous to lift one batch of methoxide, therefore several smaller batches may need to be mixed. One mixing tank for the methoxide will therefore be beneficial, but there must also be a way of checking the completeness of the mix.

These are important issues that will need to be answered before committing to any specific design. Again, it would be best to seek advice in online biodiesel chat rooms to see what others have done & then decide on which method would best suit your needs.

There should also be a method of releasing the methoxide into the processing tank in a controlled manner to avoid any hot spots or areas of incomplete conversion.

It will always be better to pump or gravity feed methoxide into the processor rather than pour & then expose the environment to methanol fumes.

Washing

In the wash tank, which wash system will be included? It is recommended that more than one system is used, therefore which two will be included? The best options would be mist washing & bubble washing, therefore will there be sufficient mechanical ventilation installed to remove the resultant methanol vapours?

Drying

Drying the biodiesel before it can be used must also be included, therefore does the design allow for the tank to be heated following the washing?

All these elements will need to be adequately designed into any successful biodiesel processor. This is true for any DIY build or for an off the shelf turn key system as every step of the process of making biodiesel must be undertaken in a controlled & safe manner, or the whole process will move from being uncontrolled to dangerous very quickly.

Therefore, ensure that EVERYTHING is safe before committing to any design, or the inclusion of anything within the design.

Also, will the design be very simple & controllable, or very complex & automated. These decisions must all be made when opting on a preferred design.

The following two pages show some DIY setups found for sale online.

Figure 41 A very basic biodiesel processor (Public Domain © 2020)

Figure 42 A less basic biodiesel processor (Public Domain © 2020)

Figure 43 A well made comprehensive setup (Public Domain © 2020)

Note in figure 43 the inclusion of a pump at the base of each tank as well as each tank being independent & fitted on castors to allow the whole system, or independent tanks to be pushed outdoors for processing.

Electrics

As the design will undoubtedly use electric, be it for pumps, timers, heaters & the like, it is important that the correct rated cabling is used & that the whole system has its own dedicated circuit/s which is fed from a consumer unit which is fitted with circuit breakers.

The services of an electrician may need to be employed to ensure the electrics comply with all current safety standards.

Storage

When it comes to storing biodiesel, there are a small number of problems & therefore rules to observe, to enable you to overcome the problems.

The first issue is created by bacteria. Due to the fact that biodiesel is a type of modified food, there are bacteria which will be happy to move in & start eating through the fuel. Therefore when storing biodiesel, it must be undertaken in such a way that it makes the environment as unwelcoming as possible for the bacteria, thereby minimising the ability of the bacteria to feed & breed on the fuel. These can therefore be seen as biological countermeasures.

As biodiesel is hygroscopic (absorbs moisture from the air), any contact of biodiesel with the air must be minimised as water in the biodiesel will not only damage the engine, but also give any bacteria more food. To overcome this, biodiesel must be stored in airtight containers & filled to a high level so that there is as little air space above the biodiesel as possible. There is always some moisture in the air, therefore by minimising contact with the air limits the ability for the biodiesel to absorb that moisture. Therefore if the biodiesel is 'dry', the bacteria will find it more difficult to live off the fuel.

Another thing that would help the bacteria to live is light. Therefore, by storing the biodiesel in the dark will also minimise the ability of the bacteria to live off the fuel. Heat too is another important factor for the bacteria. They prefer hot oil, so keeping the biodiesel as cool as possible will also inhibit the bacteria's ability to live.

Another countermeasure to be adopted in the battle against bacteria is biocide. That is poisoning the biodiesel so that it is unpalatable or toxic for the bacteria. There are commercially available biocides available that can be added to the biodiesel before it is transferred to drums or cans. It is also vitally important to only use clean dry drums & cans so as to minimise the chance of using bacterially infected storage.

Water will also hydrolyse biodiesel & it can even oxidise. Therefore these issues need to be considered too.

Hydrolysis is the term for when water breaks the bond between the alcohol & the fatty acids. This is the exact same process when which creates the FFA's in the oil & the only way to prevent this from happening is to keep the oil dry.

For oxidisation to occur, oxygen is required. When the unsaturated fatty acids oxidise, they produce a really foul smell because as the fatty acids break down into shorter & shorter chains, the acidity will increase. Therefore it is somewhat easy to detect if the biodiesel has oxidised as it will smell rancid, but it is difficult to know how long it has before it oxidises. Also, as the fatty acids polymerise & group together, the biodiesel will become more viscous. The peroxide value will also increase as free radicals will oxidise into peroxide (O_2^{2-}) & hydroperoxide (H_2O_2).

When biodiesel molecules come into contact with certain metals, heat or light, a free radical is formed. This is typically what occurs with unsaturated fatty acids with two or more double bonds. The free radical is highly reactive & will join to any oxygen molecules that it comes into contact with & form a peroxide (O_2^{2-}) molecule. When that peroxide molecule brushes against another unsaturated fatty acid molecule it creates another free radical & the peroxide molecule will become a hydroperoxide (H_2O_2) molecule, & then another free radical.

This will then continue its destructive process. The molecular process will only stop when two free radicals brush against each other & thereby damage themselves. They then form carbonic acids, aldehydes & alcohols.

Biodiesel will absorb four times as much oxygen than water; therefore reducing any contact with oxygen will greatly reduce its ability to form these free radicals. Also, it must also be kept in the dark & not allowed to contact metals, or be exposed to high temperatures. The only other defence one has to protect the longevity of the biodiesel is to use antioxidants. Antioxidants can be added so that they attack the free radicals & make them less reactive. The food industry uses this method to help stop food from spoiling. Vegetable oils have natural antioxidants, therefore the food industry have just replicated this natural preservative.

Therefore, when storing biodiesel, adding biodiesel specific biocides & antioxidants will help the longevity. By following these simple rules, the useable lifespan of the fuel will be greatly enhanced.

There is a test called the Rancimat test method, where air is passed through hot oil & deionised water. The readings from these actions are checked off a chart & it should predict the time it will take for the oil to turn rancid. In the EEC this test is covered under EN14214 & in the USA in ASTM D6751.

Design thoughts

The design & the components on any system will undoubtedly be decided by two factors. What can be afforded & what is required. Whether it is a simple system, or a complicated & automated layout, those two factors will decide what is ultimately bought or made.

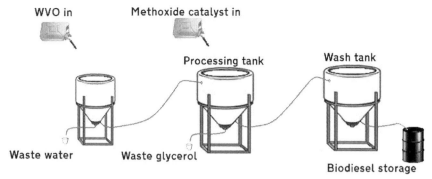

Figure 44 A minimalist layout (P Xavier © 2020)

If designing & building an effective processor seems somewhat challenging, there is a system available that may be suitable. It is called an appleseed processor & many companies & individuals have these ready to post out. Typically, they are 95% complete & only an electric supply & a few small items are needed to install & run them. It is even possible to buy plans online to make your own appleseed processor.

They are a low cost single tank solution & many people have had great success using them.

Chapter 11 – Cleaning up

There are two by-products that will need to be dealt with when making biodiesel. They are the water & the glycerol. Both contain the left over chemicals that were not fully used up in the creation of the biodiesel, therefore both these by-products will need a degree of processing to render them harmless.

Waste water

All the water (H_2O) that was used in producing the biodiesel should not be flushed down the drain. It will contain methanol, soap & small quantities of all the chemicals that were used during the biodiesel production process. These chemicals are dangerous & this is why the PPE is needed. It can not in good conscience be simply poured down the drain.

The first item to be removed will be the methanol.

Methanol recovery

If the waste water is left for a week or two outdoors, the methanol will evaporate from the water naturally. But because the methanol is the most expensive chemical used in the biodiesel process, some recover the methanol for reuse in later batches.

To do this, the waste water needs to be kept in a sealed container (to stop the methanol escaping), then this should be transferred to a still where the water can be heated & the methanol vapour tapped off & cooled. This will allow the methanol to condense back into a liquid. Methanol boils at 64.7°C (149°F), therefore the water will need to be heated to slightly higher than this temperature. 65°C (149°F) will be high enough to liberate the methanol, but low enough so that there is limited evaporation of the water.

The still should consist of a metal container that can contain a sizeable amount of water, be able to be heated without a flame & strong enough to remain stable at 65°C (149°F). The container will also need an airtight closable opening in the top in which to pour the waste water, a one way valve fitted to allow air to be sucked into the vessel & a coiled hose wrapped in wet cloth to condense the methanol vapour. The condensate should then drip into a suitable methanol container.

Some use a vacuum pump to remove the methanol vapours that sit above the water, but this is a more dangerous method as if it is an electric pump, there is a danger of sparks causing an explosion, or if it is a hand pump, then it will be hot enough to burn skin. Also, standing in the proximity of the heated methanol vapours will pose a real danger to health, therefore using a still & observing it from a distance is by far the safest method.

The still should be heated continuously until the heated water has given up the methanol.

The methanol has now been removed from the water, but soap & potassium hydroxide (KOH) will remain mixed in, therefore these will need to be tackled next.

Soap & grease recovery

Any residual soap or grease that is contained in the water will now need to be removed. If the water is left to settle for a week, then the soap & grease will naturally rise to the top of the water & then it can be removed. A quicker way would be to slowly pass the water through a grease trap.

A trap can be easily constructed with a small tank. The waste water is fed into one side & exits through the other side. The inlet is placed higher than the exit pipe & it is always beneficial to include a baffle plate down the centre to encourage the unwanted grease & soap to rise.

It will also be beneficial to drill a small vent hole in the top of the outlet pipe to prevent any siphoning occurring.

Commercial traps are available, but making one should not prove too difficult an exercise to make an effective grease trap. After the water has been circulated through the trap, the unwanted soap & grease will float to the top of the trap where they can be removed.

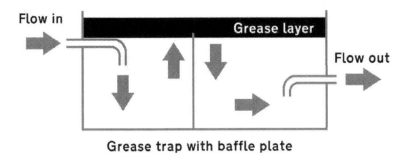

Grease trap with baffle plate

Figure 45 A simple grease trap (P Xavier © 2020)

Dissolving some magnesium sulphate ($MgSO_4$) aka epsom salts into the water will also help liberate any soap from the water.

Dealing with the potassium hydroxide

Potassium hydroxide (KOH) is an alkali, therefore it can be neutralised by simply adding a suitable acid. One such acid would be hydrochloric acid (HCl) which is yet another dangerous corrosive chemical, therefore great care must be taken & why PPE should still be worn during every stage of biodiesel production. An alternative acid that is slightly less dangerous would be acetic acid ($C_2H_4O_2$).

Care should also be taken to slowly increase the pH level of the water back up to 7 pH by carefully adding & mixing in the hydrochloric acid. When the pH has reached 7, the water can then be disposed of down a drain.

Next, the glycerol needs to be dealt with.

Waste glycerol

The glycerol ($C_3H_8O_3$) as previously stated also contains impurities. It will have soap, methanol & sodium hydroxide (NaOH) all mixed in & if left for too long, the glycerol will harden, therefore it would be sensible to process the glycerol as soon as possible. The glycerol mix is made up of approximately 49% methanol, roughly 49% glycerol & the remainder being made up of soap & the leftover catalyst, although some sources state that the mix is made up of 50% glycerol, 40% methanol & the remaining 10% is made from soap & the leftover catalyst. The exact proportions are not too important, what is important is processing this waste to make it safe. One important thing to note is that if sodium hydroxide (NaOH) is used as the catalyst for the biodiesel, the glycerol will become a solid. If potassium hydroxide (KOH) is used, then the glycerol will remain in more of a liquid state.

Methanol recovery

As was seen with the waste water, heating will remove the methanol. Again a still can be used & as methanol boils at 64.7°C (149°F), the glycerol will need to be heated to slightly higher than this temperature. 65°C will be high enough to liberate the methanol from the glycerol. All the same safety measures must be taken as was seen with the methanol recovery from water & exactly the same equipment will be needed.

Dealing with the potassium hydroxide

Again, as with the waste water, potassium hydroxide (KOH) is an alkali, therefore it can be neutralised by simply adding a suitable acid. One such acid would be hydrochloric acid (HCl) which is yet another dangerous corrosive chemical, therefore great care must be taken & why PPE should still be worn during every stage of biodiesel production. An alternative acid that is slightly less dangerous would be acetic acid ($C_2H_4O_2$). Care should also be taken to slowly increase the pH level of the glycerol to 6.88 pH by carefully adding & mixing in the hydrochloric acid. When the pH has reached 6.88 pH, the glycerol is neutral.

Measuring the pH of the glycerol will best be undertaken with an electronic pH meter, not litmus paper because glycerol can be very dark. The glycerol will coat the paper & it will then be impossible to read, let alone read with any accuracy, therefore the electronic pH meter will be the best way to accurately read the pH level.

Getting the glycerol to its natural 6.8 pH will also have other benefits.

Soap & grease recovery

When the glycerol reaches its neutral pH state, any soap will be converted to free fatty acids (FFA's), which will make the glycerol cloudy. If left to settle, two distinct layers will emerge. The topmost layer will be the FFA's & the bottom layer will be the glycerol. The FFA's can therefore now easily be removed & stored ready for mixing in with the next batch of WVO that is destined to be processed into biodiesel.

There should also be some crystals or power that has formed at the bottom. This is the sodium hydroxide & should be removed & disposed of in a responsible manner. One way to dispose of it would be to place it in a compost bin.

The glycerol at this stage should now be semi clear & much lighter in colour. If this is the case, the glycerol should be suitable for making into soap, but before soap making is looked at, there are some other options that need to be examined. Just in case the glycerol that you have hasn't quite made the desirable soap making grade.

Composting the glycerol

Glycerol is a biodegradable product, therefore it can be composted. To do this, it will need to be added along with a suitable medium such as shredded cardboard, paper, leaf litter & the like. This allows it to soak into the matter where it can then start to break down. If it is placed in neat, it will have a lower surface area & therefore be exposed to less oxygen, so be unable to break down fully. Adding to shredded cardboard & the like will allow it to be in contact with the air & also allow the bacteria to have a larger area to work with.

It will then add valuable nutrients to the soil. Also, the sodium hydroxide (NaOH) can be mixed in too, because lye is a valuable nutrient for the soil & is classed as a fertiliser in agriculture. Both the potassium hydroxide (KOH) or sodium hydroxide (NaOH) can be added to the compost as these are both fertilisers. The glycerol will therefore be safe to add to any compost bin so long as the methanol has been removed first, but beware, when the glycerol still contains potassium hydroxide (KOH) or sodium hydroxide (NaOH) it will still be caustic, therefore it will burn skin. It is therefore advisable to remove these leftover catalysts before it is used in the compost.

It is also important not to overload the compost bin & allow a few months for the glycerol to completely break down. If too much glycerol is added to the bin, it will not break down, therefore it is better to often add a little, than a lot in one go.

Also, if the glycerol is added with water (33% glycerol, 66% water), it will act as an activator & greatly speed up the anaerobic digestion & composting process.

Burning the glycerol

Excess glycerol can also be used as a fuel in log burners, but not in barbeques. The reason for this is that when it is burnt, it can release acrolein fumes (when burnt at temperatures above 280°C 536°F). These acrolein (C_3H_4O) fumes are not only highly toxic, but are also a very strong irritant for the skin, eyes & nasal passages, therefore it should never be burnt in an open fire, only in fires that have chimneys to remove any possible toxic fumes.

The best way to prepare the glycerol for burning is to mix it with sawdust or shredded cardboard. The sawdust/glycerol mix can then be mixed & compressed to form a solid block. Aim for a ratio of 40% sawdust, 60% glycerol & compress as much as possible to form a brick. The higher the compression, the better the result will be. There are easy to use briquette presses available online for sale for less than £20 that makes this a simple process. These briquettes can then be used in a suitable fire & one brick should burn for between 45 minutes to 1 hour. A supply of these bricks will be sufficient to heat a home which has a suitable fire, but it is not a smokeless fuel. It can burn with thick, acrid smoke if it is burnt at too high a temperature.

The risk from burning glycerol is minimal so long as the temperature is kept below 280°C (536°F). All nicotine vapers burn glycerol as it is the chemical used as a base to carry the nicotine & flavouring. These vaping machines are designed to operate between 190 – 235°C (374 – 455°F), therefore so long as the briquettes are not burnt at too high a temperature, they will be safe to use too.

Commercial incinerators are designed to burn glycerol at temperatures in excess of 1,000°C (1,832°F), which is so hot, it vaporises any acrolein that can be present. This is also the same process that the army uses to destroy dangerous chemicals in old & unwanted chemical weapons, such as mustard gas shells left over from WWII.

Dust suppression with glycerol

Glycerol is a humectant chemical. This means that it absorbs moisture, therefore if a glycerol/water mix is sprayed on gravel or a dirt roads, it stops dust becoming airborne. In many parts of the world, it is used exactly in this manner.

Due to ever more stringent pollution controls, large building sites have to suppress dust that originates from their work sites. This usually entails constantly spraying water onto the unmade roadways & hosing down vehicles before they exit the site. Therefore spraying a glycerol/water mix on the unmade roadways will remove the problem of dust escaping the site. This is only applicable to building sites during the summer months as wet winter weather keeps the ground sufficiently wet to stop any dust forming.

It is also ideal (providing all the methanol, potassium hydroxide (KOH) & sodium hydroxide (NaOH) has been removed) to spray in arenas where dust is a problem; a horse riding school for example. It will stop dust being kicked up by the horses & is suitable for use indoors as well as outdoors.

There are therefore business opportunities to be made, selling the cleaned glycerol to anyone who needs to control their dust.

Soap making with glycerol

If after removing the methanol, soap & leftover catalyst (potassium hydroxide (KOH) or sodium hydroxide (NaOH)) the leftover glycerol should be clear & transparent, then it should be pure enough to use to make soap.

As a rule of thumb, 38.5g of sodium hydroxide (NaOH) will be used for every litre of glycerol (after removing all the impurities) & also 250ml of water for every litre of glycerol.

Simply, heat the glycerol in a stainless steel cooking pot such as a maslin pan to 66°C (150°F), just to be double sure that all the residual methanol has been removed.

In a separate stainless steel cooking pot heat the water to 40°C (104°F), then add the correct amount of sodium hydroxide (NaOH) & mix until it is completely dissolved into the water. Pour the water/NaOH mix into the heated glycerol & continue to heat the complete mix for 10 minutes whilst continuously stirring, then stirring very gently for a further 10 minutes. After the 20 minutes, draw a spoon through the mixture & if a line remains visible for a few seconds, it is ready to decant. If not, continue to heat & stir for 10 minutes & then repeat the spoon test.

When the mix is ready, it can be decanted into soap moulds or Tupperware type containers to a depth of 50mm, then set to one side to cool & cover with a suitably sized piece of cardboard placed on top.

After 24 hours, the soap can be released from the moulds. If Tupperware containers have been used, the block can be cut with a sharp knife into suitably sized bars. These then need to be left for a minimum of 7 days before they can be used (these bars will lighten over the 7 days as they continue to solidify & they will sweat too as they dry) & they are then best stored in either airtight containers (such as Tupperware) or freezer bags & kept in a cool place.

Greaseproof paper can be used to stop the bars sticking to each other whilst being stored. If the soap bar stings when it is used, it will need to be given more time to cure.

If soap making is to become a permanent fixture for the excess glycerol, it may be worth purchasing a soap kettle, soap moulds & a selection of soap dyes & essential oils. If the results look professional, a small scale business opportunity could be investigated.

Chapter 12 – Winter biodiesel

Biodiesel does not have the same qualities as petrochemical diesel; therefore there are some instances when biodiesel will need to be slightly enhanced to achieve the level of performance that you will expect from the finished biodiesel product. One of those qualities where it falls short is winter motoring.

Winter motoring

Petrochemical diesel which is bought at the forecourts in winter months is winterised fuel. This is a different mix to that which they sell during summer months as in winter, there are additives premixed into the fuel. These additives enhance the petrochemical diesel in two ways. Firstly, they reduce the 'Cold Filter Plugging Point' (CFPP). This is the name for the point when the fuel starts to crystallise when at a low temperature. The temperature that this happens is dependant on what the fuel is made from & how it is made, but always makes the fuel appear cloudy, hence why the CFPP is also known as the 'Cloud Point'. As these crystals grow, they will become visible when they are four times larger than the wavelength of visible light & this is why the fuel becomes cloudy, but as the temperature continues to drop, these crystals will continue to grow in size & will eventually get to a point where they become big enough to block a fuel filter.

If the temperature continues to drop even further, these crystals will grow enough to turn the fuel from a liquid to a solid. The temperature at which this happens is called the 'Pour Point', or 'Gel Point'. When this stage has been reached, the fuel can no longer be pumped by the fuel pump as the fuel is no longer a liquid, it has become a solid.

There are additives available for petrochemical diesel which lowers the temperature at which this cloud point occurs & therefore lowers the gel point too. It is these additives that are already mixed in to the petrochemical diesel that is sold over the winter months.

Unfortunately, biodiesel also suffers from clouding & gelling too, but unfortunately there is nobody standing by in a white lab coat who will add anything to proof your biodiesel against low winter temperatures & the majority of these additives available are designed solely for petrochemical diesel, so they do not work too well in biodiesel. Therefore winterising biodiesel is something that must be undertaken if temperatures are to be expected below -5°C (23°F). Fortunately, there are a few tricks available for biodiesel that will help.

First, here's how to measure for the cloud point & gel point without having to send biodiesel to a petrochemical lab for expensive testing.

Measuring the cloud point & gel point

It should be noted that every batch of biodiesel that is produced will be slightly different & every batch of biodiesel made from different oils may be drastically different. Therefore if four 50 litre batches were made from 200 litre drum of WVO, each of the four batches would give similar results, but four 50 litre batches made from four 50 litre batches from different sources could net completely different results. This is because all the oils used to make SVO each have different properties, as was seen in figure 32.

Hydrogenated & partially hydrogenated oil will act just like naturally saturated fats. Also, just like the saturated fats, the higher the concentration of hydrogenated or saturated fat, the higher the cloud point & gel points will be. Figure 45 lists the cloud point temperatures for the ten example oils that were first examined back in chapter 4.

Therefore to avoid potential problems with the biodiesel, selecting the best oil based composition of the SVO or WVO will reduce the need for winterising the biodiesel.

CP Oil	Temp
Peanut	6
Rapeseed	-3.3
Hemp	-4
Safflower	-2
Coconut	0
Oil palm	13
Soybean	0.9
Sunflower	3.4
Jatropha	10.2
Corn	11.5

However, if the choice of SVO is limited, or if there is no choice with the WVO, then steps can be taken to resolve the cold weather issues.

Therefore assuming, the oil content used to make the biodiesel was unknown, the biodiesel's cloud point will need to be obtained. To do this, you will need the following items:

Figure 46 Cloud point temperatures of oils (P Xavier © 2020)

- Empty jam jar with hole in lid big enough to insert a thermometer
- Thermometer
- Fridge freezer

Fill the jam jar to approximately half full with the biodiesel that is to be tested & place the thermometer into the jam jar, so that it is standing in the biodiesel. The hole in the top of the jam jar will allow the thermometer to stand upright. Place the jam jar in the fridge & keep checking the sample every few minutes.

When the biodiesel sample starts to become cloudy, make a note of the temperature as this is the cloud point temperature of the biodiesel.

The fridge only reduces the temperature to 4°C (39°F), so the sample should then be placed in the freezer & checked every two minutes for the cloud point temperature.

When the sample starts to gel, make a note of the temperature as this will be the gel point temperature.

Although these two tests will not enable you to tell what oil the biodiesel was made from, you will know at which temperatures the cloud point & the gel point will be. But to get ready for the following winter, if it is possible, it would be best to choose an oil with a low saturated fat content. This is the key to making good cold weather biodiesel. Biodiesel created from rapeseed (aka canola) has the lowest cloud point of the ten oils that were examined because rapeseed oil is very low in saturated fats. Therefore, if it is possible, use only rapeseed, hemp or safflower oils.

If using these oils is not an option, then there are other countermeasures that can be employed.

Reducing the cloud point

This can be achieved by simply heating the biodiesel, then cooling it until it starts to cloud. Keep it at the cloud point temperature & allow time for the cloudy (crystals) to fall to the bottom. Separate the clear oil from the top & set it aside.

The cloudy biodiesel can be stored & labelled as 'summer biodiesel', whilst the clear oil can be used as 'winter biodiesel'. It is the saturated fats in the biodiesel that started to crystallise, so by removing those crystals, the remainder of biodiesel will cloud at a lower temperature. The winter biodiesel will still start to cloud at approximately -5°C (23°F), therefore if lower temperatures are to be anticipated, then an alternative solution will need to be adopted. Also, it may be impractical to chill & possibly freeze a large volume of biodiesel in one batch; therefore it may not be an option for everyone.

Block heater

Some diesel engines are equipped with block heaters. These are sometimes fitted as standard to some diesel powered vehicles.

It will be worth checking the specifications of your vehicle to see whether one is present or not. If there is a block heater present, it may need to be connected to a suitable electric cable & then left plugged in overnight. If not, there are numerous retro fit block heaters available to purchase online.

The most common type available uses an electric heating element either in or on the engine block. In cold areas such as Northern Canada, Northern Europe & Northern Russia, vehicles can be seen driving around with electrical plugs protruding from the front grilles. This is where the electric cable is plugged into whilst the vehicle is parked up. In Mongolia, they rely on lighting small fires on the floor under their engine, about an hour or so before they intend to drive, but this Mongolian method would not be recommended.

These heaters range in price from £20 up to around £500. The very cheap end of the market would not be recommended.

Battery heater

The internal chemistry in wet & gel batteries can freeze in cold weather & therefore make starting the vehicle engine more difficult because the battery will be underperforming, or worst case scenario, not provide power at all. Therefore a battery warmer & battery insulation may be key to operating a diesel (biodiesel) powered vehicle in winter months.

Even if the battery was wrapped in a thermal blanket (the thin silver paper type), it will perform much better overall.

An electrical heater with thermal wrap can be obtained for approximately £50. As with the block warmer, this type of heater will need to be plugged in to an electrical supply & left switched on when the vehicle is parked up.

Tesla electric cars are now fitted with battery warmers as standard which are powered when the vehicle is plugged in & charging. This allows the batteries to be warm enough to perform well, even in cold weather.

Engine heater

This is another type of electrical heater, but these connect to a coolant hose. This is then heated & the heat will circulate the coolant around the engine. There are also oil pan heaters that connect either to the sump or oil pan & these warm the oil. Another variant is the electric element type that replaces the dipstick. Again any of these will need to be plugged into a suitable electric source when the vehicle is parked overnight.

Fuel tank heater

There are also electrical pads available to heat the fuel tank. There are the mains powered type available & also 12v models designed to be powered from the vehicle battery. These are very similar to the barrel heaters that were seen in chapter 10, but much smaller.

Fuel filter heater

Similar to the tank heating pads, there are also smaller fuel filter pads available to heat the fuel filter. These are also available in mains & 12v variants.

Fuel additives

As mentioned earlier, there are additives that can be added directly to the fuel to chemically lower the cloud point & gel points, but these are typically designed only to work with petrochemical diesel & therefore do not work too well with biodiesel, but there are a small number available.

Further measures

There is also the option of the 'Night Heater' that is typically found in HGV's & used to heat the drivers cab whilst the engine is switched off. This is an ingenious little heater which is very quiet in operation as to operate it just has diesel dripped onto a glow plug. This burns & the heat warms the air in a neighbouring chamber where the warm air is then blown out. This air can obviously be routed wherever it is required; therefore it can be routed into the engine bay to keep the engine warm. The exhaust gasses are vented out via a small exhaust. There is no reason why biodiesel would not work with these heaters; therefore the fuel supply could be tapped off the fuel line.

These can be expensive to fit, but are excellent heaters. New wave custom conversions sell one of these types of heaters[55].

---

[55] https://www.newwavecustomconversions.co.uk/air-heaters/ - 31/07/2020

Chapter 13 – What went wrong?

As making biodiesel relies on multiple steps, which needs exact quantities of chemicals which need to meet strict quality standards, which are all heated, handled & separated in a precise manner, the chances of mistakes & errors occurring somewhere in the process can be rather high. Therefore, it is important to know how to resolve any issues that could occur.

My biodiesel doesn't look right. What's gone wrong?

Biodiesel should be clear enough to read a book through, be anywhere from a dark golden colour to clear & be much thinner than the oil that was used to make the biodiesel from.

Emulsion looking like mayonnaise

If it looks like mayonnaise (creamy & pale yellow), then an emulsion has been created. This can take the form of swirling in the mix, a distinct layer, it can take the form of a solid or liquid & can even affect the entire batch within the tank.

This has occurred because the glycerol, soap & methanol has been mixed far too vigorously & have combined to form an emulsion. The presence of the soap in the mix does not help as it always makes the chances of emulsification a greater probability & is therefore the most likely cause, but it can also be caused by adding water (for the wash) far too vigorously, or for too long. It could be caused by too many bubbles, or even by having bubbles that are far too big.

To minimise the chance of creating what looks like mayonnaise, ensure that ALL the glycerol is removed from the biodiesel before washing. Also, if sodium hydroxide (NaOH) was used as the catalyst, then drain off the glycerol several times. The first time after 3 hours, then at 3 hour intervals until the mix is ready to progress to the next stage of processing.

The first wash should always be a static water wash, then after the water has had time to react, remove it & only then move onto a gentle mist wash. The mist washing should be repeated (changing the water regularly). Only after it appears milky white, can more vigorous washing techniques be used.

When bubble washing, ensure the bubbles are not too large or that the bubbling is not too violent. Also, as bubble washing progresses, the wash water will become ever more saturated & therefore increasing the frequency for changing the wash water will also help, or you may be bubbling saturated soapy bubbles through the biodiesel which will also cause an emulsion to form.

If the emulsion has formed in the mix, leave it sit for several days & it should eventually clear itself & then it will be possible to drain it away. If after a week, if it still does not clear, try one of the following (first on list is most potent, last is least potent):

- Add a little glycerol
- Leave it sit for another week
- Gently heat the mix
- Add a little salt

To add glycerol, you must have some stored somewhere. If this is the case, then it must have the methanol still in it. It is therefore worth keeping about 20 litres in storage in case it is ever needed for breaking emulsions.

Begin by gently heating the emulsified biodiesel mix & the methanol containing glycerol to approximately 34°C (93°F).

Then gently add the heated glycerol to the mix about 10% at a time & gently stir into the mix.

Continue to add the heated glycerol up to a maximum of 50%. The emulsion should break & when it does, the colour will return to normal & the glycerol will then settle out of the mix. It should therefore be left to settle & after 2 hours, drain out the glycerol.

The next best alternative to adding glycerol is to just wait for another week. If the emulsion disappears, remove the glycerol & then wash, if not heat a few litres of water to around 50°C (122°F) & add to the mix with a very fine watering can & then gently stirring the mix for a minute or two. After it has settled, remove & dispose of the water & repeat the process of adding heated water & spraying on top several times.

Adding salt should only be undertaken as the last resort as it will make the toxic emulsion even more toxic & also make the mix even more corrosive to metals. Therefore after removing the emulsion, the whole tank will need to be thoroughly cleaned to remove all traces of the salt. This works as the water molecules are more attracted to the salt molecules than to the soap molecules, so in theory should drop the soap & jump to the salt molecules. When this happens, the soap should start to be removed from the emulsion & allow it to settle out.

Thick gel looking like pudding or jelly

If the mix has formed into a thick gel which has the appearance of a pudding or jelly, then there was either too much water or too much catalyst present. This could be due to inaccurate scales being used to measure the catalyst ingredients or other miscalculations. It should be important to note that if a scale that measures to 1g & rounds down, 1.9g could still read as 1g & therefore the weight could be 90% out & enough to completely spoil the mix. Also, the presence of cleaning products contaminating the oil can also cause a gel.

There is no method to recover the mix if it turns to gel, but to help prevent any mixes turning into gel, observe the following rules.

Undertake several titration tests using oil that has been well mixed. Never just test the oil that is at the top. Only use accurate scales & measure everything carefully. Always double check the calculations & if possible ask A N Other to check the results. Test the oil for the presence of water before undertaking the biodiesel processing & always dewater the oil. NEVER process oil that has a high FFA content.

Oil & glycerol not separating out

If a distinct layer of glycerol can not be seen below the biodiesel (or far less than the volume of methanol that was added), then it is likely that either the oil failed to react, or hardly reacted with the catalyst.

There is an abundance of monoglyceride & diglyceride molecules in the mix. These occur when only one or two of the FFA's break from the glycerol molecule. These can not be seen by the naked eye & will not wash out from the mix & can even help the mix to emulsify. They will even clog the injectors in the engine, corrode the engine & also cause NOx emissions.

They can be caused by having an excess of water in the oil which could cause soap which inhibits the chemical processes, using oil that titrates too high, insufficient agitation of the mix, insufficient quantities of catalyst &/or methanol being used & also by a failure to heat the mix to a high enough temperature.

It can be resolved by ensuring that the catalyst is measured correctly & by ensuring that the mix is stirred adequately & for long enough. It is always better to mix slightly longer than is necessary, rather than shorter than is required.

Ensure that the mix is kept at around 50°C (122°F) during the entire reaction process, but ensure that the temperature does not reach 64.7°C (148°F) as this is the temperature at which methanol boils. Use 22% methanol if 20% is not enough to fuel the reaction. Obtaining a complete reaction is key to successfully making biodiesel.

Reprocessing the batch is the only way to recover any batches that suffer from incomplete reactions, but during the second processing, add less methanol & catalyst. There's no way of knowing exactly how far the original reaction went, or exactly how much to add. Experience will be the only guide here in this scenario. Alternatively, either dilute it (perhaps 5% at a time) into the next 20 batches, or experiment with it perhaps 1 litre at a time until you have worked out exactly how much catalyst & methanol is needed to complete the reaction.

Completing the biodiesel steps without completing the transesterification process & then using the defective biodiesel will result in damaging your engine.

There's a soapy layer in the batch

Soap can form as a whitish layer between the glycerol & biodiesel. It can also appear as a foamy layer on top of the biodiesel. This only happens when an excess of the catalyst has been used, water is in the oil or poor quality oil was used. Some soap is to be expected, but should always be kept to a minimum as it will lead to emulsions forming.

Great care must be taken with the catalyst. Undertake several titration tests using oil that has been well mixed. Never just test the oil that is at the top. Only use accurate scales & measure everything carefully. Always double check the calculations & if possible ask A N Other to check the results. Test the oil for the presence of water before undertaking the biodiesel processing & always dewater the oil.

After removing any soapy scum that appears on the surface, undertake a static wash before moving onto mist washing. Then, ensure that any soap layer has been removed by the mist washes before moving onto the bubble washes (which should be gentle).

Solids in the biodiesel

Any solids that occur in the biodiesel will wear the injectors, pumps & damage the engine, therefore it is important that the biodiesel is adequately sieved with a 5 or 10 micron sieve before it is used. Even if you believe the biodiesel is clean & free from solid contaminants, it isn't. Even particles that are too small to see with the naked eye can cause engine problems, therefore ALWAYS SIEVE THE BIODIESEL.

Acidic biodiesel

If the biodiesel is too acidic, it is most likely caused by using oil that contained too much FFA's which is most likely caused by either inaccurate titrations or incomplete processing. If the biodiesel has been stored & then found to be acidic, then it has possibly been exposed to heat, light or air that have caused free radicals to form in the fuel.

Therefore, it is important to undertake several titration tests using oil that has been well mixed. Never just test the oil that is at the top. Only use accurate scales & measure everything carefully. Always double check the calculations & if possible ask A N Other to check the results. Test the oil for the presence of water before undertaking the biodiesel processing & always dewater the oil.

If it is to be stored, ensure it is in an airtight container, kept in the dark & in the cool. Make certain that there is a minimal air gap above the fuel, add biocide & antioxidants.

To check if it is a storage issue or not, undertake a reprocess test to see if the transesterification process has completed or not.

How do I know if the biodiesel is clean enough to use in my car?

So long as the biodiesel has been passed through a 5 or 10 micron sieve then it should be fine, but if you really need to check, using a petrochemical hydrometer, check the specific gravity of the biodiesel & it should have a reading of between 880 & 900.

The fuel filter in the engine keeps getting blocked. What's wrong with the biodiesel to cause this?

Petrochemical diesel is a very dirty fuel & therefore leaves oily & greasy deposits behind. Therefore any engine that has had petrochemical diesel used as a fuel has also had the fuel tank, the fuel lines & everything that has come into contact with petrochemical diesel coated with greasy deposits. These deposits will also attract any dirt that has found its way into the system, therefore basically every engine has these dirty, greasy deposits lining the fuel system.

Biodiesel (if made correctly) will be very clean & will also start to strip all those greasy deposits that were left from the petrochemical diesel. Therefore particles of the grease will float in the biodiesel & travel through the system, only to get caught in the fuel filter. This is perfectly normal. It is happening because the fuel system is being cleaned by the biodiesel.

If after two or three months there is a noticeable loss of power, change the fuel filter, or better still. Install a cheap in-line filter somewhere in the fuel line before the fuel line gets to the fuel filter. If the cheap in-line filter gets blocked, just change it.

It will be best to carry spare fuel filters in the vehicle & have the tools & knowledge of how to change them because if the filter becomes blocked, the engine will not run until the filter is changed.

I'm loosing engine power when I use biodiesel. Is this normal?

There is a 5% drop in power when switching from petrochemical diesel to biodiesel (B100). However, it is unlikely that this drop in power will be too noticeable, therefore the first thing to do is change the fuel filter. If it makes little difference, change the air filter & check the fuel lines for bulges, splits & cracks. If any of the lines are defective, replace with a suitable fluroelastomer hose such as Viton.

It is also possible that the biodiesel has liberated a lot of the greasy deposits that were left from the petrochemical diesel. If possible, take a sample of fuel from the fuel tank to check for any greasy sludge. If there is sludge present, then the fuel tank will need to be removed & cleaned.

The diesel injectors may also be becoming coked. If cleaning these are beyond the skills you enjoy, then any competent mechanic will be able to check the injectors & clean them.

The warning lights have come on in my vehicle when I used the biodiesel. Is this normal?

Some of the engine sensors in the diesel engine are designed to read the oxygen level, but as biodiesel is an oxygenated fuel, it contains far more oxygen than petrochemical diesel, therefore in some cars, the sensor senses that there is more oxygen & therefore incorrectly assumes there is water in the fuel.

Then the computer turns on the 'water in fuel' warning light, or the 'service' light.

These are only warnings & can therefore be ignored, but if you must check, it is best to open the drain valve at the bottom of the fuel filter to check for the presence of water in the fuel.

Can I add petrol to my biodiesel to help it run in winter?

Absolutely not! Adding petrol to biodiesel **will just damage the engine**.

Can I just use old engine oil to fuel my diesel engine?

It is possible to use old engine oil to fuel your engine, but only if you have a two tank system & that you filter the oil completely. However, it will still be dirty & there will be very fine particles of metal in the oil & these will damage your engine. Also, the engine will suffer from coking, the emissions will never pass any pollution control checks & it will be belching black smoke filled with NOx. It would therefore only be suitable for large robust engines used off road, such as a 360 excavator.

How pure is the glycerol byproduct?

The glycerol is a very crude product which will contain methanol, catalyst & water, which therefore needs careful processing before being able to make soap.

If there are doubts about the quality of the glycerol after processing it, then it can be used as a degreasing soap which can be used on garage & industrial floors. If the glycerol is as described in chapter 11, then it should be fine to use as soap.

Chapter 14 – Various tests

There are numerous tests that can be undertaken on the biodiesel to check that it is good enough to put in your engine & also tests to check the oil that you intend to use to make biodiesel is a good oil. Many of those tests are listed here.

The level of conversion test

The level of conversion test is used to determine whether or not the oil has completed its transformation into biodiesel. To do this, the biodiesel will be tested to see how soluble it is in methanol.

SVO will not dissolve into methanol, but instead forms two distinct phases of methanol & SVO. Monoglycerides & diglycerides will partially dissolve in methanol. Unlike pure biodiesel the methanol will become saturated with both the monoglycerides & diglycerides. When this saturation point occurs, it will be unable to dissolve any more. This test will therefore show whether the biodiesel has more vegetable oil, diglycerides & monoglycerides than the methanol can readily absorb.

To perform the test, heat a small quantity of biodiesel & methanol to 20°C (68°F). Then add 4ml of the heated biodiesel to a small glass graduated cylinder, a volumetric pipette or a burette along with 36ml of the heated methanol. The methanol must be pure & unused, not recycled or recovered from an earlier biodiesel mix. The biodiesel/methanol mix should then be mixed & allowed to settle for 15 minutes.

If after the 15 minutes there is a distinct layer on the bottom, then the biodiesel has failed the test. If the biodiesel has been absorbed by the methanol or if it looks a little cloudy, then these both pass the test.

The cloud point & gel point test

To obtain the biodiesel's cloud point you will need the following items:

- Empty jam jar with hole in lid big enough to insert a thermometer
- Thermometer
- Fridge freezer

Fill the jam jar to approximately half full with the biodiesel that is to be tested & place the thermometer into the jam jar, so that it is sitting in the biodiesel. The hole in the top of the jam jar is to allow the thermometer to stand upright. Place the jam jar in the fridge & keep checking the sample every few minutes.

When the biodiesel sample starts to become cloudy, make a note of the temperature as this is the cloud point temperature of the biodiesel.

The fridge only reduces the temperature to 4°C (39°F), so the sample should be placed in the freezer & checked every two minutes for the cloud point temperature.

When the sample starts to gel, make a note of the temperature as this will be the gel point temperature.

The HMPE filtration test

When biodiesel becomes cold, a white grease like substance or small white flakes will appear to be suspended in the biodiesel. These suspended particles are known as High Melting Point Esters (HMPE's). They have a high melting point & do not merge back into the biodiesel at room temperature, but they will be reabsorbed at high enough temperatures.

To undertake the test, you will need the following items:

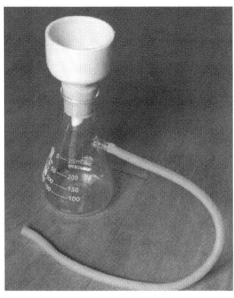

- 500ml Buckner flask
- Vacuum pump
- Vacuum tubing
- Funnel with 0.7 micron filter
- 300ml biodiesel

Construct the apparatus as shown in figure 47. Chill the biodiesel to 4°C (39°F) the temperature that a domestic fridge should be set at, for 16 hours. Then remove from fridge & allow to naturally return to room temperature.

Figure 47 Buckner flask apparatus (Public Domain © 2020)

After the temperature of the biodiesel has returned to room temperature without any intervention, attach the tubing to the vacuum pump & place the filter into the funnel section of the Buckner flask. Pour the room temperature biodiesel into the funnel & time how long the 300ml sample takes to go through the 0.7 micron filter.

If the whole of the 300ml sample passes through the filter in less than 360 seconds, then the sample has passed the test. If there is anything caught in the filter, then it is the HMPE's.

The FFA test

It is possible to measure the FFA content in oil with 3M Oil Quality Test Strips. They are available in two versions. Standard range (1004) & low range (1005). The standard range strips will measure FFA content between 2 – 7%, whilst the low range strips measure between 1 – 2.5%. They are used in exactly the same way as litmus paper, but here the coloured sections indicate the FFA content, not the pH.

The methanol test - purity

Methanol will absorb water from the air over time, therefore every time the bottle is opened, it will absorb a little more water. Even when the bottle is closed, the air in the top of the bottle or container will have water in it & the methanol will absorb that water & it is the water content of the methanol that can cause problems in biodiesel production.

If a batch of biodiesel does not go to plan or an emulsion forms, it could be caused by too much water being present, therefore this test can be used to determine if the methanol has been tainted with water. If the methanol is less than 95% pure, then it can cause problems in the mix.

The easiest way to test the methanol is to use the petrochemical hydrometer. This is best used in a graduated cylinder. The specific gravity is the ratio of a substance's density in relation to the density of water. Therefore if a liquid is lighter than water, then the specific gravity will be lower than 1. Anything that has a specific gravity greater than 1 are therefore heavier than water & the density is the mass of a substance divided by its volume. The test is also temperature sensitive as everything expands & contracts with temperature, therefore to achieve accurate results, the temperature must be recorded along with the density.

The specific gravity of methanol at 20°C is 0.7913 & as the methanol absorbs water, the specific gravity will increase until it reaches a maximum of 1.000 at 20°C.

$$\text{Purity of methanol} = \frac{(1 - \text{specific gravity})}{0.2087}$$

Therefore if the specific gravity reads 0.823 at 20°C, then using the formula, the purity of the methanol would be 84.8%.When purchasing a petrochemical hydrometer, it will arrive with a chart so that the result can be adjusted for different temperatures.

The pHLip test

A test has been developed to check biodiesel for common problems, they are, the quality of manufacture & the pHLip test will give a result on the quality as a percentile. The second problem is poor storage & this results in the fuel becoming oxidised. Both of these will cause fuel filter clogging & also damage to the engine.

It is impossible to achieve a 100% conversion rate when making biodiesel as the process starts to reverse immediately. The process is continually going backwards & forwards in the processing tank, which is why extra methanol is used because it tips the balance in favour of the required reaction. As was seen earlier, when fuel ages, it breaks down & oxidation occurs. The acid number can be determined easily using glassware & undertaking a simple experiment, but a free & total glycerol test requires a 'gas chromatograph' to be accurate, but this is an expensive piece of equipment, therefore the pHLip test was developed.

The test is contained in a small plastic bottle that is half filled with the test chemicals. The biodiesel is added & the bottle shook, then left for a time before the results become visible.

Further information can be obtained from the pHLip website[56].

The separation test

This test will help reveal when it is time to stop water washing the biodiesel. Simply transfer approximately 100ml of biodiesel from the wash tank into a laboratory beaker. If removed during a wash cycle, it will have the appearance of orange juice, then place the beaker in a warm spot & allow it to settle. Time how long the sample takes to become clear. If it takes longer than 10 minutes, then more washing is required.

---

[56] http://www.phliptest.com/index.html - 03/08/2020

If it takes less than 10 minutes to clear, then the water washing is complete.

The take away titration test

This test is so called as it can be used behind the take away when foraging for oil to use. It is a basically a 'field' titration test. Determine the maximum titration number that you require from the oil, then create a test kit in small sample bottles at that number.

Just add 10ml of isopropyl alcohol in a small sample bottle then add three drops of phenolphthalein. Create neutral result using your KOH standard solution & to that add 1ml of the KOH standard solution. This will give a titration value of 1, therefore for 1.5, add a further 0.5ml etc, until the desired value has been reached.

At the take away, using a pipette, just add one ml of the oil, shake it up & read off the results. That way, you can select only the oil that you want & reject the oil you do not want.

Tests for water in WVO

Water in WVO & biodiesel can exist in three ways; dissolved in, emulsified & free. When dissolved, the water molecules are floating around, suspended in the WVO. It is too small to see as there is not enough of it in one place to be big enough to see. Therefore the WVO can hold an amount of water & still remain totally clear. Also, WVO can hold up to five times more water than SVO in this way.

When the WVO becomes saturated with water but it keeps absorbing still more, the water forms a fog in the WVO & it is known as an emulsion. This emulsion will make the WVO appear cloudy.

To add even more water to the WVO will increase the water molecules even further until they form water droplets in the oil. This will be something similar to rain drops, & then these rain drops will drop out of the bottom as the WVO has become too saturated to hold any more water.

The maximum amount of water permissible in finished biodiesel is 0.05%. This is the upper limit recommended by the biodiesel community & also by the EEC under EN14214 as the presence of water in the biodiesel will damage an engine. It is therefore important to be able to test for the presence of water & if found to be greater than 0.05%, remove it from the biodiesel. There are therefore a few ways of measuring for the water content.

The weigh test

If you are in possession of an accurate set of scales, it will be possible to find the water quantity using those scales. Using this method, the water content can be found down to 1000 – 2000 parts per million.

The following equipment will be needed:

- A weighing scale accurate to 0.1g
- A borosilicate conical flask
- A thermometer
- Electric hotplate
- Stainless steel cooking pot or maslin pan
- A borosilicate stirring rod
- 1kg of WVO

Weigh the maslin pan & make a note of the weight (weight a). Then add the 1kg of WVO to the pan & weigh it again (weight b), then place the pan on the hotplate & heat the WVO to 120°C (248°F), ensuring that it is constantly stirred to stop the water vaporising & scalding the surrounding area with boiling oil.

When the temperature of the WVO has reached 120°C (248°F), remove from the heat & weigh the oil filled pan again (weight c).

Now it's just a simple mathematical calculation to find out how much water the WVO contains.

Deduct weight c from weight b & record the result as the 'water weight'. Deduct weight a from weight b & record the result as the 'oil weight'. Divide the oil weight by the water weight & multiply by 100. This will give the water content of the WVO as a percentage.

As an example, if it assumed that the maslin pan weighs 200g (weight a) & with 1 litre of oil added it will weigh 1,200g (weight b). Therefore the oil weight is 1000g. After the WVO has been heated, it is weighed again & reads 1,180g (weight c). Therefore the water weight is 980g. Dividing water weight (980g) by the oil weight (1000g) gives a value of 0.98. Multiplied by 100 will give a percentage value of 98%. Therefore the sample contains 2% water.

By multiplying this figure by 10,000 will give the parts per million value, therefore 2 multiplied by 10,000 will give a water content value of 20,000 parts per million.

The sandy brae water test kit

In this test, a small sample of water is mixed with a small sample of calcium hydride (CaH_2). The calcium hydride reacts with the water & hydrogen gas (H) is liberated. This process is undertaken in a small flask & the liberation of the hydrogen creates a positive pressure inside the flask which can be read from the dial on the lid.

Figure 48 Sandy Brae test flask (Sandy Brae Laboratories 2020)

Full details for the kit is available from the Sandy Brae Laboratories website[57].

Titration of WVO

Titration is used to find out exactly how much extra sodium hydroxide (NaOH) or potassium hydroxide (KOH) is needed to be added to the oil to create biodiesel.

See titration, titration, titration on page 156 for details on this test.

---

[57] http://www.sandybrae.com/Water_Test_Kit.html - 04/08/2020

Chapter 15 – Alternative biodiesel recipes

There are numerous recipes that can be found on the internet. If the example recipe that was used earlier in the book is too difficult, or does not work for you, then try one of the other recipes. Some of those alternatives are listed here.

Single stage recipes

Here are a number of single stage recipes that can be used. Most people opt for a single stage recipe, but one size does not fit all as everyone will compromise somewhere. For some, those compromises will often be based on cost or quality, but for others the compromise could be based on their equipment, their processor design, brew time or even safety.

Despite all the compromises, the standard biodiesel recipe should be thought of the gold standard. The liquid gold standard. If you have a newer type diesel engine, then it will run best on quality biodiesel, but a very old engine may be happy to run on suboptimal biodiesel. However, try the standard first.

The standard biodiesel recipe

- Methanol – use 22% by volume of the WVO
- Potassium hydroxide (KOH) – (8g + titration) x litres of WVO, or
- Sodium hydroxide (NaOH) – (5g + titration) x litres of WVO

If the titration of the WVO is high, then opt for the potassium hydroxide (KOH), not the sodium hydroxide (NaOH).

Temperature = 55 – 60°C (131 – 140°F)

Follow the instructions in chapter 9 – the first batch for all the processing instructions, but substitute the volumes of materials listed for this recipe & if it is for a large batch in a processor (reactor), then during washing, first undertake a static wash, then mist wash several times before bubble washing.

The high yield biodiesel recipe

This recipe should only be used if the biodiesel is to be used in an IDI engine as it results in a mix of biodiesel & WVO. Therefore if the biodiesel from this recipe is used in a common rail or direct injection engine, then **it will cause damage**.

- Methanol – use 15% by volume of the WVO
- Potassium hydroxide (KOH) = (5.5g + titration) x litres of WVO, or
- Sodium hydroxide (NaOH) = (3.5g + titration) x litres of WVO

Temperature = 55 – 60°C (131 – 140°F)

Whilst the oil is heating, make the methoxide, then when up to temperature add the methoxide mix. Agitate the whole mixture for one hour & leave to settle for 24 hours.

If separation has occurred, remove the glycerol. There is no wash stage so the mixture contains methanol, glycerol & catalysts, therefore it is an extremely impure mix of biodiesel, therefore **only use it in IDI engines**.

The high titration adjustment factor recipe

If the titration level of the WVO is extremely high, then multiplying the titration by a numeric factor may prove helpful.

It is identical to the standard recipe, but multiply the titration number by a factor of 1.15 & then add it to the 5g base to derive the amount of sodium hydroxide (NaOH) to use.

Obviously, all of the above recipes can be slightly adjusted to best suit the equipment that you have available. For instance, if your mixer is a little underpowered, then add a little time to make up for the shortfall.

Two stage base recipes

Here are a small number of two stage recipes that can also be tried out. A two stage process will help achieve better results if the standard recipe is not yielding the desired quality biodiesel.

The 80 – 20 recipe

This method is just the same as the standard, except after the methoxide is mixed, it is split into two portions. One being 80%, the other 20%. This is done because transesterification is a reversible chemical process & therefore an excess of methanol is needed to ensure complete conversion. When there is more methanol present than glycerol, it tips the balance in favour of the transesterification direction that is desired. That is turning WVO to biodiesel. Therefore, this helps to improve the reaction by shifting the ratio of glycerol to methanol even further in favour of the methanol to achieve a complete reaction.

The base base biodiesel recipe

If quality is not an issue but you wish to reduce the methanol input, then this may help achieve that. It aims to use the minimum amount of methanol & catalyst during the first stage, then it is as normal for the second stage.

First stage use:

- 140ml of Methanol
- 3.5g Sodium hydroxide (NaOH) + (titration + 1.15), or
- 4.9g Potassium hydroxide (KOH) + (titration + 1.15)

For the second stage use:

- 23ml of Methanol
- 1.5g Sodium hydroxide (NaOH), or
- 2.1g Potassium hydroxide (KOH)

It will take a minimum of 14% methanol to achieve separation when a thorough mixing method is employed. If there is any trouble achieving adequate separation during the first stage, then slightly increase the amount of methanol used in the first stage ever so slightly. Then, in order to adjust for the fact that higher titration oils may not separate every time, the titration value is multiplied by 1.15 for the first stage.

The second stage should not need to be changed as the aim is to have converted all the FFA's during the first stage.

The process is exactly as was described as the 80 – 20 method, except the quantities listed above should be used.

The zero titration two stage recipe

The concept behind this recipe is to use a set amount of catalyst in the first stage, then quantify the level of conversion using the 'level of conversion test' (see chapter 14) & then setting the exact amount of catalyst needed for the second stage from those results. This recipe has the advantage of removing the need for a titration test, but just replaces it with a 'level of conversion test', so overall, there is no real advantage.

Therefore, once the percentage of unconverted oil after the first stage is known, only then can the amount of ingredients for the second stage be assembled. Also, as there will be no FFA's present in the second stage, there will be no need for any extra catalyst. Therefore, you will only need to adjust the base catalyst down to account for the already converted oil. So if it is assumed that the test resulted in 15% unconverted oil, it would be a simple case of multiplying the base catalyst of 5g/L by 0.15 (or 15%) & the batch size in litres to derive the amount of catalyst that will be needed for the second stage.

Therefore to begin, 5g of sodium hydroxide (NaOH) or 7g of potassium hydroxide (KOH) will be used as the first amount of catalyst. Therefore, the first stage would call for 80% of the total methanol & 5g per litre for the catalyst.

To demonstrate this, assume that a 50 litre batch is to be made using 20% methanol with a 15% fallout after stage 1.

Stage 1:

- 250g of Sodium hydroxide (NaOH), or
- 350g of Potassium hydroxide (KOH)
- 10l of Methanol x 0.80 = 8 litres of methanol

Stage 2:

Percentage of fallout, multiplied by the base, multiplied by the volume = 0.15 x 2.5g/l x 50 = 18.75g of NaOH

10l of Methanol x 0.20 = 2 litres of methanol

Two stage acid recipes

There is also a two stage acid recipe that can also be undertaken, but it should only be undertaken by experienced biodiesel processors.

The FATTA recipe

This is used in many commercial biodiesel scenarios & it is used on oil that titrates between 2 – 6 with sodium hydroxide (NaOH). It can be used to make biodiesel from oils that have a high & very high FFA content.

The FATTA process will create water as a byproduct & as the concentration of water increases in the mix, the whole reaction will slow. Therefore, to counteract this, the acid stage is often undertaken in steps & the resultant water is drained between each step.

Stage 1 – the acid catalyst to esterify the FFA's:

- 1 litre WVO
- 100ml Methanol (CH_3OH)
- 1 – 2ml Sulphuric acid (H_2SO_4) (at 98% pure)

React at 45 – 55°C (113 – 131°F) until the titration falls to a minimum level somewhere near 2.6ml, which could take up to three hours. If more acid is added, it will just raise the titration end point.

Stage 2 – the alkali catalyst (transesterification):

- 100ml Methanol (CH_3OH)
- 6.4g Sodium hydroxide (NaOH)

The 6.4g is derived from 3.5g for the catalyst + 2.9g for acid neutralisation (assuming a titration of 2.6ml was achieved).

React at 45 – 55°C (113 – 131°F) for one hour, then allow the mix to settle & separate.

Chapter 16 – Glossary of terms

There has been a high amount of technical & chemical terms in this book & therefore there may be a huge amount of words which the reader may be unfamiliar with. This is also true of many biodiesel websites & discussion boards. Hence this glossary should help as these terms can be quickly looked up to check on the definition of these words.

ABS – acrylonitrile butadiene styrene (C_8H_8-C_4H_6-$C_3H_3N)_n$, this is a thermoplastic which is used in a wide variety of products. It can be damaged by solvents, but has some resistance certain chemicals (except strong oxidisers). It has a very good strength & has good resistance to cracking.

Acetone – (C_3H_6O) aka propanone, dimethyl ketone, 2-propanone, propan-2-one & beta-ketopropane. It is a solvent used in soap tests & is typically found in fingernail polish remover. It is the simplest form of a ketone.

Agglomerator – this is an apparatus designed to combine things together. Ball shaped sweets are tumbled in an agglomerator & this gives them their shape. In the case of biodiesel, it is used to separate water from the fuel, so that the two do not mix.

Alkali solution – when an alkali is dissolved in water, the solution is known as an alkali solution. In biodiesel production, the alkali is used in the titration & is made by mixing the catalyst (either potassium hydroxide (KOH) or sodium hydroxide (NaOH)) in water.

Anhydrous – this literally means 'no water'. In chemistry, anything that does not contain water is anhydrous. In biodiesel manufacturing, the transesterification process must be without water & is therefore anhydrous.

Anaerobic digestion – this is a sequence of processes where micro organisms break down biodegradable material in the absence of oxygen, whilst composting uses oxygen.

Antioxidant – this is a substance that inhibits oxidation. Oxidation is a chemical reaction that produces free radicals, which in turn cause damage. Therefore, including an antioxidant in biodiesel will help with long term storage.

Atomisation – this is when the fuel is sprayed as a fine mist by the injectors into the combustion chamber of a diesel engine.

Atomise – see atomisation.

Azeotrope – this is when two or more liquids are combined & their proportions can not be altered or changed by distillation. When the azeotrope is boiled, the vapour has the same proportions of constituents as the un-boiled mixture.

Base – a base is not an acid, but can react with acids & neutralise them. Bases are usually metal oxides or metal hydroxides such as with potassium hydroxide (KOH) & sodium hydroxide (NaOH).

Biodiesel – this is a biofuel created from natural sources to replicate the characteristics of petrochemical diesel.

Biodiesel recipe – this is a set of instructions for making or cooking something, in this case biodiesel.

Bonded glycerol – propane (C_3H_8) which has three OH groups added is called glycerol. These OH's love to bond with hydrogen & therefore bond with FFA's. Therefore the molecules of the WVO are made from a glycerol molecule 'bonded' to three chains of FFA's.

Bromophenol blue – this is a chemical ($C_{19}H_{10}Br_4O_5S$) that can be used as a pH indicator in biodiesel production.

Bubble wash – this is a method of removing water, methanol & the impurities it is carrying by bubbling air in water which is contained in a separation layer below the biodiesel. As the air bubble rises through the biodiesel it has a thin layer of water surrounding it which attracts methanol & its impurities to it as it travels upwards. The bubble then bursts at the top of the tank & the water then sinks to the bottom of the tank, collecting more methanol & impurities on its way down.

Calcium stearate – this is a chemical soap ($C_{36}H_{70}CaO_4$) containing one calcium atom & two fatty acid chains. It is the main component of soap scum & is created when biodiesel is washed in hard water. It is insoluble in water & is used as a food additive under the number E470 & it is used to coat smarties.

Canola – this is a trademarked name from the 'Rapeseed Association of Canada'. The name Canola originates from 'Can' from Canada & 'Ola' meaning oil. It is now a generic term for edible varieties of rapeseed (*brassica napus*) that were developed to contain less toxic erucic acid.

Carbonation – this is when something combines with carbon dioxide (CO_2). It is what gives fizzy drinks their fizz, but in biodiesel production, if sodium hydroxide (NaOH) is left exposed to the air, it will combine with carbon dioxide (CO_2) & form sodium carbonate (Na_2CO_3), rendering it useless for biodiesel production.

Catalyst – this is a substance that speeds up the rate of a chemical reaction & is not consumed by the reaction. It also remains unchanged chemically & in mass by the end of the reaction. In biodiesel production, sodium hydroxide (NaOH) or potassium hydroxide (KOH) are the catalysts for the transesterification reaction.

Caustic – this is any chemical that has the ability to burn, corrode, dissolve or eat away by chemical action living tissue or some other substance.

Cetane number – this is an indicator to the speed of & the compression needed for combustion to occur in diesel type fuels. It is the diesel equivalent to the octane rating in petrol.

Cloud point – the temperature that solid wax crystals start to form in biodiesel when the temperature drops. These can cause blocked fuel filters, fuel lines etc..

Coking – a term used for when carbon deposits start to build up on the fuel injection nozzles & the piston which is caused by impurities in the fuel such as glycerol or a high level of monoglycerides, diglycerides & triglycerides due to unsuccessful reaction in the biodiesel processing.

CPVC – chlorinated polyvinyl chloride ($C_9H_{11}Cl_7$) is a common thermoplastic pipe with improved heat resistance & flexibility properties than PVC, but melts when in contact with biodiesel.

Deliquescence – this is the chemical process by which a substance absorbs moisture from the atmosphere until it dissolves in the absorbed water & then forms a solution. Both sodium hydroxide (NaOH) & potassium hydroxide (KOH) will absorb water from the atmosphere until the point when they eventually become just a puddle of water.

Diglycerides – SVO is made from glycerol molecules with three attached chains of Fatty Acid Methyl Esters (FAME's). When one of the FAME's break off, the glycerol molecule still has two FAME's attached, so it is called **diglyceride**. Break off a further FAME & only one is left attached to the glycerol molecule so it is now called monoglyceride. The presence of diglyceride & monoglyceride in the biodiesel will ultimately damage the engine, so it is important to liberate all three of the FAME's & then separate them from the unwanted glycerol.

Emulsion – this is a mixture of biodiesel & water which can be a milky white colour that can sometimes be difficult to separate.

An emulsion can be caused by one of several ways. Excessive agitation during the wash can cause the soap in the biodiesel to mix with water, or an incompletely reacted biodiesel mix which contains unwanted FFA's can hydrolyse causing an emulsion when water is added.

Ester – these are chemical compounds derived from an acid (organic or inorganic) in which at least one OH (hydroxyl) group is replaced with an O (alkoxy) group. For example, esters can be made from an acid in which at least one hydroxyl molecule is replaced with a molecule from the alkyl group. Formic acid (CH_2O_2) can therefore be transformed into the most simple of the esters. The result of which is a highly flammable, colourless liquid with a delicate odour called methyl methanoate ($C_2H_4O_2$) aka methyl formate which is a methyl ester.

Esterification – this is the chemical term for a chemical reaction which two reactants (typically an acid & an alcohol) form an ester.

Ethanol – this is a simple alcohol that is also called ethyl alcohol, grain alcohol, drinking alcohol, spirits & also alcohol. It has the formula C_2H_6O.

Ethyl ester – an ethyl ester is an ester of ethanol.

Exothermic – when a chemical reaction happens, energy is transferred either to or from the surroundings. When the energy is transferred to the surroundings it is called an exothermic reaction & during this reaction the surrounding temperature increases. Examples of exothermic reactions are combustion, some oxidation reactions & some neutralisation reactions.

FAME – fatty acid methyl esters are a type of fatty acid ester which is derived by transesterification of fats by methanol & the molecules that make up biodiesel are primarily FAME's.

Feedstock – this is any renewable, biological material that can be used as a fuel, or converted into a different form of fuel or energy product. In biodiesel manufacturing, the feedstock is the oil that is converted into biodiesel. WVO, or SVO.

Free fatty acid (FFA) – SVO & WVO is made from glycerol molecules with three attached chains of Fatty Acid Methyl Esters (FAME's). When the oil is heated, the molecules become damaged & some of the FAME's break free. These free floating chains are the free fatty acids (FFA's).

Free glycerol – this is the amount of glycerol that is suspended in the biodiesel before it is washed. Any left in the biodiesel after washing will damage the engine.

Free radical – this is any molecule that has at least one unpaired electron. They are highly reactive as they will interact & break other molecules thereby creating more free radicals. If biodiesel is stored for too long, the fatty acids will polymerise & group together, then the biodiesel will become more viscous. The peroxide value will also increase as free radicals oxidise into peroxide (O_2^{2-}) & hydroperoxide (H_2O_2). When the peroxide molecule brushes against another unsaturated fatty acid molecule it creates another free radical & the peroxide molecule will become a hydroperoxide molecule, then another free radical is created. This chain reaction will spoil the biodiesel.

Flashpoint – this is the minimum temperature at which a chemical gives off enough vapour to ignite, providing there is a sufficient concentration of the vapour.

Fluorocarbon – this is a chemical compound that contains carbon & fluorine & is used in lubricants, refrigerants, non stick coatings & aerosol propellants. It is also used in many resins & plastics.

Gel point – the temperature that biodiesel begins to solidify into a gel when the temperature drops. This will cause blocked broken fuel pumps, fuel filters, fuel lines etc..

Glyceride – this is an ester formed from glycerol & fatty acids. There are three chains of fatty acid attached to one glycerol molecule.

Glycerol – this is a simple chemical compound that is colourless, odourless & viscous which is sweet tasting & non toxic. It has the formula $C_3H_8O_3$.

HDPE – high-density polyethylene is a thermoplastic polymer made from ethylene. It is high strength, therefore used in HDPE pipes & also bottles.

Hydrocarbons – this is an organic chemical compound that is comprised solely of hydrogen & carbon atoms. These are naturally occurring compounds that form the basis of fossil fuels such as crude oil, natural gas & coal.

Hydrogenated oil – this oil (aka trans fat) has been chemically altered to remove any double bonding in the molecules. The purpose behind hydrogenation is to lengthen the shelf life of cooking oil, but it is then difficult to make fuel from hydrogenated oil as the whole process makes the oil become more solid.

Hydrophilic – this means 'water loving', & only hydrophilic chemicals will dissolve into water. Anything that will dissolve into water is therefore hydrophilic.

Hygroscopic – anything that is hygroscopic will readily attract & absorb moisture from its surroundings either by absorption or adsorption. Sodium hydroxide (NaOH) & potassium hydroxide (KOH) are both hygroscopic.

Immiscible – the inability of two or more liquids to be mixed together. For instance, water will readily mix with other liquids, but will not mix with oil, therefore oil & water are immiscible.

Indicator solution – an indicator is added to a solution to determine if a substance is acid, neutral or alkali. It does this by changing colour to represent the pH level. Phenolphthalein ($C_{20}H_{14}O_4$) & bromophenol blue ($C_{19}H_{10}Br_4O_5S$) are both indicators.

Isopropyl alcohol – this is an alcohol also known as 2-propanol, *sec*-propyl alcohol, IPA & isopropanol (C_3H_8O). It is colourless, highly flammable & has a strong odour that smells of a mixture of ethanol & acetone. It is used in titrating WVO.

Kinematic viscosity – this is the measure of a fluids internal resistance to flow under gravity which is measured over time.

KOH – see potassium hydroxide

LDPE – low density polyethylene is a thermoplastic made from the monomer ethylene. It has excellent resistance to chemicals therefore is typically used for processing tanks in biodiesel production.

Linolenic acid – this is an essential fatty acid ($C_{18}H_{30}O_2$) belonging to the omega-3 group & is highly concentrated in certain plant oils & is a chemical compound found in biodiesel.

Litmus paper – litmus is a water soluble mixture of different dyes that are extracted from lichens. When absorbed on to paper strips it forms a pH indicator & changes colour in the presence of acids.

LLDPE – linear low density polyethylene is a type of polyethylene that has a high tensile strength along with good impact & puncture resistance.

Lubricity – the measure of the reduction in friction or wear by a lubricant. Biodiesel has high lubricity.

Lye – an alternative name for sodium hydroxide (NaOH).

Magnesium stearate – this is a chemical compound $Mg(C_{18}H_{35}O_2)_2$ which is a soap that is made from stearic acid (a fatty acid made from a chain of 18 carbon atoms $C_{18}H_{36}O_2$) & magnesium (Mg). It can sometimes form in biodiesel if it is washed in hard water.

Methanol – (CH_3OH) aka methyl, methyl alcohol & wood alcohol is a very dangerous, volatile chemical. When making biodiesel, technical grade methanol is used which is 98 – 99% pure. At room temperature it takes the form of a clear, light, volatile, colourless, highly flammable liquid that has a distinctive odour similar to drinking alcohol. In biodiesel production it is used for the transesterification of WVO into biodiesel. Methanol is highly toxic & will burn with an invisible flame.

Methyl ester – this is a type of fatty acid ester which is derived from SVO or WVO by transesterification of the fats using methanol.

Methoxide – a mixture of methanol (CH_3OH) & either potassium hydroxide (KOH) or sodium hydroxide (NaOH) that is used in biodiesel production.

Miscibility – the ability & property of two liquids (or substances) that will combine to form a homogenous solution. This is the ability to dissolve into each other at any concentration.

Mist wash – this is a method of removing water soluble impurities that are present in biodiesel. It is achieved by spraying a fine mist of water over the top of the biodiesel. These small droplets of water will then sink down through the biodiesel, absorbing the impurities as it sinks.

Molecular sieve – this is a material with pores of a uniform size that are small enough to block entry by large molecules but big enough to allow the passage of small molecules.

Monoglyceride – SVO is made from glycerol molecules with three attached chains of Fatty Acid Methyl Esters (FAME's). When one of the FAME's break off, the glycerol molecule still has two FAME's attached, so it is called diglyceride. Break off a further FAME & only one is left attached to the glycerol molecule so it is now called **monoglyceride**. The presence of diglyceride & monoglyceride in the biodiesel will ultimately damage the engine, so it is important to liberate all three of the FAME's & then separate them from the unwanted glycerol.

Monounsaturated fat – these are fats that contain only one double bond in the molecular chain.

NaOH – see sodium hydroxide

Nitrile – this is any organic compound that has a -C≡N functional group. It can take many forms such as methyl cyanoacrylate ($C_5H_5NO_2$) which is used in super glue & nitrile rubber which is used in latex-free gloves. Nitrile rubber is also used in the automotive industry due to its ability to resist fuels & oils.

Partially hydrogenated oil - this oil (aka trans fat) has been chemically altered to remove any double bonding in the molecules. The purpose behind hydrogenation is to lengthen the shelf life of cooking oil, but it is then difficult to make fuel from hydrogenated oil as the whole process makes the oil become more solid. Partially hydrogenated oil is exactly the same, but is semi-solid & still contains many trans fats.

pH – is the measure of acidity or alkalinity of a aqueous solution & represented on a sliding scale. Lower values are acidic & higher values are more alkaline. At room temperature, water is neutral at pH 7.

pH test strip – see litmus paper.

Phenol red – a solution of phenol red ($C_{19}H_{14}O_5S$) is used as a pH indicator & is expressed between yellow to red. It can be used as an indicating solution when titrating WVO.

Phenolphthalein – ($C_{20}H_{14}O_4$) is a white or yellowish crystalline which is a weak acid & used as an indicator in chemistry & also as a laxative in medicine. It is colourless below pH 8.0 & becomes pink at pH 9.3. It then becomes a deep red above pH 9.6. It can be used as an indicating solution when titrating WVO.

Polythene – aka polyethylene ($C_2H_4)_n$ is the worlds most common plastic & has excellent chemical resistance properties.

Polyunsaturated fat – these are fats where the hydrocarbon chain has two or more carbon double bonds. All fatty acids that contain double bonds are called unsaturated fats. Unsaturated fats have a higher gel point, but are more unstable & the bonds tend to break down more easily. Oils that contain polyunsaturated fats include safflower, sunflower, corn & soybean.

Potassium hydroxide – aka caustic potash, lye, potash lye, potassia & potassium hydrate (KOH) is a base that can be used to make biodiesel. It dissolves in methanol easier than sodium hydroxide (NaOH) & the soap byproduct is liquid rather than the solid soaps formed when using sodium hydroxide (NaOH).

Potassium stearate - this is a chemical compound ($C_{18}H_{35}KO_2$) which is a soap that is made from stearic acid (a fatty acid made from a chain of 18 carbon atoms $C_{18}H_{36}O_2$) & potassium (Mg). It is formed when using potassium hydroxide (KOH) as the catalyst for making biodiesel.

Polypropylene – aka polypropene ($C_3H_6)_n$ is a thermoplastic polymer which has excellent chemical resistance properties, so it is widely used in biodiesel production.

Processor – this is the tank or vessel that holds the WVO & methoxide while it chemically reacts & changes into biodiesel. It is also called a reactor.

Polyvinyl chloride – aka PVC is a synthetic plastic polymer that is resistant to acids, salts, bases, fats, fuels & alcohols.

Polyvinylidene fluoride – aka PVDF $(C_2H_3F)_n$ is a non-reactive thermoplastic fluoropolymer that is resistant to chemicals & also heat.

Rapeseed – see canola.

Renewable diesel – fuel made from animal derived fats as opposed to biodiesel which is made from plant derived fats.

Rancimat test – this is a test used to estimate the time that it will take for biodiesel to turn rancid.

Reactor – see processor.

Saponification – this is a chemical process that involves converting fats, oils or lipids into soap & alcohol by the action of heat using an alkali such as sodium hydroxide (NaOH) or potassium hydroxide (KOH).

Saturated fat – a fat is saturated when each of the carbon atoms in a fatty acid molecular chain has two associated hydrogen atoms. It is therefore saturated with hydrogen. So they are fatty acids without double bonds. The lower the saturated fat content, the lower the gel point for any oil made from it. Saturated fats are typically solid at room temperature.

Smoke point – this is the temperature that SVO or WVO will start to smoke. The higher the value of FFA's, the lower the temperature it will start to smoke.

Soap test – this is the test that is used to determine amount of soap present in oil or in unwashed biodiesel. It is a titration test that uses hydrochloric acid (HCl) solution as a reagent, bromophenol blue $(C_{19}H_{10}Br_4O_5S)$ as an indicator (.04% in water), & acetone (C_3H_6O) or isopropyl alcohol (C_3H_8O) as a solvent.

Sodium hydroxide - (NaOH) is a base that is used to make biodiesel.

Sodium methoxide – (CH_3ONa) is a chemical compound which is formed by the deprotonation of methanol & is used as a reagent. It is also a dangerously caustic base. It can also be used as catalyst for transesterification when making biodiesel.

Sodium stearate – ($C_{18}H_{35}NaO_2$) is the sodium salt of stearic acid & is the world's most common type of soap. It is made from one stearic acid molecule (a fatty acid made from a chain of 18 carbon atoms $C_{18}H_{36}O_2$) & one fatty acid molecule. It will form in biodiesel when sodium hydroxide (NaOH) is used as the catalyst.

Static wash – this is a method of removing water & the impurities it is carrying by placing water in the same tank as the biodiesel without any agitation. This cleansing will occur at the boundary of the water & biodiesel as the water, methanol & its suspended impurities migrate downwards into the water.

Stir wash – this is a method of removing water & the impurities it is carrying by placing water in the same tank as the biodiesel & agitating. This cleansing will occur as the water & biodiesel mix & the water will attract the suspended impurities before migrating downwards to rest at the bottom of the tank. If the mix is agitated too vigorously, it will cause an emulsion to form.

SVO – straight vegetable oil.

Titration – this is a chemical test used in biodiesel production to resolve how much catalyst is needed to turn all the FFA's contained in the WVO to soap. It uses either sodium hydroxide (NaOH) or potassium hydroxide (KOH) as the catalyst.

Trans fatty acids – these are a type of unsaturated fat, where hydrogenated & partially hydrogenated oils have been chemically altered to remove the double bonding to lengthen the shelf life of the cooking oil, but it is then difficult to make fuels from these oils as the process makes the oil more solid.

Transesterification – base-catalysed transesterification is the name for the process where lipids (fats & oils) are reacted with an alcohol (typically methanol or ethanol) which in turn forms biodiesel & glycerol. This is possible because vegetable oils & animal fats are predominantly made from triglycerides.

Triglyceride – the process of base-catalysed transesterification involves triglyceride molecules. SVO is made from glycerol molecules with three attached chains of Fatty Acid Methyl Esters (FAME's) aka **triglyceride**. When one of the FAME's break off, the glycerol molecule still has two FAME's attached, so it is called diglyceride. Break off a further FAME & only one is left attached to the glycerol molecule so it is now called monoglyceride. The presence of diglyceride & monoglyceride in the biodiesel will ultimately damage the engine, so it is important to liberate all three of the FAME's & then separate them from the unwanted glycerol.

Unsaturated fat – this is oil that is more commonly found in plant & fish based material & can be monounsaturated (containing only one double bond in the molecular chain) or polyunsaturated (containing two or more double bonds in the molecular chain). Unsaturated fat is a liquid at room temperature.

Viscosity – this is a measure of a fluids resistance to flow. It is perceived as thickness in the fluid. Water is very thin, so it has a low viscosity, whilst honey is thick, therefore has a high viscosity. Biodiesel should have a similar viscosity to petrochemical diesel, which is far lower than the viscosity of SVO or WVO.

Viton – this is a brand name for a synthetic rubber & fluoropolymer elastomer which will resist corrosion by biodiesel, therefore fuel lines can be made from Viton.

WVO – waste vegetable oil.

Chapter 17 – European biodiesel

EN14214 details the requirements of biodiesel are as follows:

Ester content minimum of 96.5% as EN14103

Density at 15°C minimum of 860kg/m^3 & maximum of 900kg/m^3 as EN ISO3675 & EN ISO12185

Viscosity at 40°C minimum of 3.5mm^2 & maximum of 5.0mm^2 as EN ISO310

Flashpoint as ≥101°C as ISO / CD 3679

Sulphur content <10mg/Kg

Cetane number minimum of 51 as EN ISO5165

Sulphated ash content maximum 0.02% as ISO3987

Water content maximum 500mg/Kg as EN ISO12937

Total contamination maximum 24mg/Kg as EN12662

Copper strip corrosion (3 hours at 50 °C) must be class 1 as per EN ISO2160

Oxidation stability 110°C – 6 hours as EN14112

Acid value maximum of 0.5mg KOH/g as EN14104

Iodine value maximum of 120 as per EN14111

Linolenic acid methyl ester maximum 12% as EN14103

Polyunsaturated (≥4 double bonds) methyl esters maximum of 1%

Methanol content maximum of 0.2% as EN14110

Methanol content maximum 0.8% as EN14105

Monoglyceride content maximum 0.8% as EN14105

Diglyceride content maximum 0.2% as EN14105

Triglyceride content maximum 0.2% as EN14105

Free glycerol maximum 0.02% as EN14105 & EN14106

Total glycerol maximum 0.25% as EN14105

Alkaline metals (Na + K) maximum of 5mg/Kg as EN14108 & EN14109

Phosphorus content maximum 10mg/Kg as EN14107

Any biodiesel to be sold in the EEC must meet these requirements.

Table of illustrations

The cover illustration is copyright to P Xavier (2020) & the rear illustration is copyright to Vanessa Thompson (2019).

Figure 1 Extractable energy from various fuels per kilo (P Xavier © 2019) 14

Figure 2 Diesels first experimental engine 1893 (Imotorhead64 © 2008) from 'Rudolf Diesel: Die Entstehung des Dieselmotors. Springer, Berlin 1913. ISBN 978-3-642-64940-0. p. 11, Fig 3' 16

Figure 3 Citroën Rosalie (Arnaud 25 © 2009) 17

Figure 4 Peugeot Citroën XUD engine (Public Domain © 2020) 18

Figure 5 The Diesel 4 stroke cycle (P Xavier © 2020)........... 21

Figure 6 Common fuel rail with attached injectors (Public Domain © 2020) 25

Figure 7 A diesel turbocharger (Public Domain © 2020) 26

Figure 8 Exhaust gas flow (P Xavier © 2020)...................... 30

Figure 9 Cutaway of a blocked DPF (P Xavier © 2020) 30

Figure 10 Typical diesel CAT layout (P Xavier © 2020) 32

Figure 11 Carbon chains (P Xavier © 2020)......................... 36

Figure 12 Fractional distillation (P Xavier © 2020) 37

Figure 13 A water molecule - H2O (P Xavier © 2020).......... 44

Figure 14 Hydrocarbon chains (P Xavier © 2020)................. 45

Figure 15 Auto-ignition temperatures of hydrocarbons (P Xavier © 2020) 46

Figure 16 Bonds (P Xavier © 2020) 47

Figure 17 Simple alcohol molecules (P Xavier © 2020) 49

Figure 18 Simple acid molecules (P Xavier © 2020) 50

Figure 19 Simple ester molecules (P Xavier © 2020) 51

Figure 20 Arachis hypogaea (Public Domain© 1887) 53

Figure 21 Brassica napus (Public Domain © 1887) 54

Figure 22 Cannabis sativa (Public Domain © 1887) 55

Figure 23 Carthamus tinctorius (Public Domain © 1887) 56

Figure 24 Cocos nucifera (Public Domain © 1887) 57

Figure 25 Elaeis guineensis (Public Domain © 1887) 58

Figure 26 Glycine max (Public Domain © 1804) 59

Figure 27 Helianthus annus (Public Domain © 1859) 60

Figure 28 Jatropha curcas (Public Domain © 1863) 61

Figure 29 Zea mays (Public Domain © 1897) 62

Figure 30 Diesel exhaust emissions (P Xavier © 2020) 69

Figure 31 Lengths of molecular chains expressed as percentages (P Xavier © 2020) 73

Figure 32 Breakdown of oils expressed as percentages (P Xavier © 2020) 74

Figure 33 Euro standard chart (P Xavier © 2020) 77

Figure 34 UK oil consumption (P Xavier © 2020)................... 84

Figure 35 Elsbett engine (Elsbett 1992)............................... 90

Figure 36 Base-catalysed transesterification of triglyceride (P Xavier © 2020) 92

Figure 37 Post base-catalysed transesterification of triglyceride (P Xavier © 2020)..................................... 92

Figure 38 Base-catalysed transesterification (P Xavier © 2020) 93

Figure 39 Thermal image of a pot on stove (Public Domain © 2020) 139

Figure 40 Flexible oil drum heater (Public Domain © 2020) 192

Figure 41 A very basic biodiesel processor (Public Domain © 2020) 196

Figure 42 A less basic biodiesel processor (Public Domain © 2020) 196

Figure 43 A well made comprehensive setup (Public Domain © 2020).. 197

Figure 44 A minimalist layout (P Xavier © 2020)................ 201

Figure 45 A simple grease trap (P Xavier © 2020)............. 204

Figure 46 Cloud point temperatures of oils (P Xavier © 2020) 214

Figure 47 Buckner flask apparatus (Public Domain © 2020) 231

Figure 48 Sandy Brae test flask (Sandy Brae Laboratories 2020) 236

Index

1,2,3-propanetriol......... *See* glycerol

1,2,3-trihydroxypropane*See* glycerol

2-propanol .... *See* isopropyl alcohol

2-propanone... *See* acetone

3M Oil Quality Test Strips 231

ABS 244

acetaldehyde............... 106

acetic acid 50, 51, 116, 118, 204, 206

acetone 106, 244, 251, 255

acetylene....................... 47

acrolein........................ 208

AdBlue.... 32, 33, 78, 79, 80

    AUS32....................... 32

    Bluedef...................... 79

    BlueTec...................... 79

adiabatic compression ... 15

agglomerator................ 244

alcohol. 48, 49, 64, 91, 126, 127, 128, 143, 146, 152, 157, 199, 248, 252, 257

aldehyde..... 68, 72, 82, 199

Alfa Romeo

    156 19

algae 65, 66

algal oil .... *See* algae

alkali 204, 206, 243, 244, 251, 255

alkali solution................ 244

alkaline 259

alkane........... 44, 47, 48, 49

alkene................ 47, 48, 49

alkoxy 248

alkyl 50, 248

alkyne 47, 48

alternator 86, 87

aluminium . 88, 99, 101, 103

aluminium oxide 28

ammonia........... 78, 99, 101

ammonium salt....... 99, 101

anaerobic digestion 208, 245

anhydrous 244

antioxidants .. 200, 224, 245

appleseed.................... 201

Arachis hypogaea .... *See* peanut

AR-AFFF 96, 108

Argentina 59

asphalt.......... 34, 35, 38, 46

atmospheric pressure.... 26, 27

atomisation 245

Audi 55

   100 TDI...................... 19

   R10 TDI LMP1............ 19

Austin

   A60 Cambridge 17

Australia 56, 65

azeotrope 245

bacteria... 39, 117, 198, 207

battery warmer 216

Belgium 66, 91

Bentley 17

beta-ketopropane .... See acetone

biocide 198, 200, 224

biodegradable.. 65, 83, 207, 245

biomass 60, 65, 66

block heater................. 216

BMW............................. 55

Borgward

   Hansa 17

boron 48

borosilicate .. 142, 143, 144, 154, 157, 159, 163, 164, 235

brass...................... 88, 191

Brassica napus............. See rapeseed

brassicaceae 54

Brazil........................ 59, 61

bromophenol blue 111, 112, 245, 251, 255

bund........................ 96, 107

butane................ 37, 39, 45

calcium 246

calcium hydride 236

calcium stearate 246

Cambodia 58

camshaft....................... 24

Canada 97, 99, 102, 104, 108, 114, 117, 246

cannabis sativa... See hemp

canola .. 54, 63, 74, 75, 215, 246, 255

carbon.... 44, 48, 57, 59, 63, 64, 65, 67, 70, 71, 73, 78, 82, 247, 249, 250, 252, 254, 255, 256

carbon dioxide ... 28, 31, 65, 68, 69, 70, 71, 76, 82, 83, 104, 127, 246

carbon monoxide..... 28, 31, 68, 70, 76, 81

carbonation................... 246

carbonic acid 199

carbonyl 49

Carnot's rule 22

Carthamus tinctorius ....See safflower

CAT 28, 30, 31, 32, 81

catalyst .. 28, 31, 78, 79, 91, 162, 205, 210, 220, 221,

222, 223, 227, 239, 240, 241, 242, 243, 244, 246, 254, 256

caustic 246, 256

caustic potash .... *See* potassium hydroxide

caustic soda .... *See* sodium hydroxide

centipoise 148

Central America 58

cetane number 247, 259

CFPP 212

China 13, 59

chlorine 106

Chrysler 55

CI 15, 16

Citroën 17, 18, 89

   BX 18

   Rosalie 17

cloud point ... 212, 213, 214, 215, 217, 230, 247

CN 39, 63

cocaine 114

coconut 57, 72

Cocos nucifera *See* coconut

coking 247

combustion chamber 26, 86, 245

common rail. 19, 23, 24, 25, 239

compost 206, 207

compressor 26, 27

con-rod 20

copper 88, 259

cordierite 29, 30, 31

corn 61, 62, 65, 72, 73, 254

CPR 98, 100, 102, 104, 109, 111, 112, 115, 118

CPVC 247

crankshaft 20, 27

crude oil. 34, 35, 36, 37, 38, 40, 45, 46, 83, 84, 85, 250

cycloalkanes 35

cylinder .. 15, 19, 20, 21, 22, 23, 24, 26, 29, 33, 77, 78, 229, 232

decane 45

DEF 32

deforestation 57, 58

deionised water 32, 153, 200

deliquescence 247

Denso 24

DEP 72

Department for Transport 41

de-watering .. 155, 156, 171

diaphragm 88

diffusion 27

diglyceride ... 152, 181, 184, 222, 229, 247, 253, 257, 259

dimethyl ketone .... *See* acetone

dinitrogen 69

dioxygen 69, 71

direct injection ... 19, 23, 25, 239

distillation 34

DOC 31, 32, 33, 79, 80

docosane 45

dodecane 45

dotriacontane 46

double bond. 49, 52, 73, 89, 199, 250, 253, 254, 255, 257, 259

DPF .28, 29, 30, 31, 33, 77, 80

    filter regeneration 28

    parked regeneration 29

DPM 72

drupe 57

dual fuel 85

East Asia 59

ECU 24, 28, 32, 33, 77

EEA 41

EEC .14, 18, 38, 39, 40, 41, 42, 54, 55, 57, 61, 62, 63, 65, 66, 76, 84, 96, 108, 114, 117, 200, 235, 259

EGR 77, 78, 80

    valve 77

Egypt 58

Elaeis guineensis .... *See* oil palm

electric car .... *See* electric vehicle

electric vehicles .. 13, 14, 15

electron 44, 47, 249

Elko .... *See* Elsbett engine

Elsbett engine 90

emulsification *See* emulsion

emulsion 162, 180, 181, 219, 220, 221, 232, 234, 247, 248, 256

engine heater 217

engine knocking 38

Environment Agency ... 174, 177

Environmental Protection Agency 41

enzyme 88

EPA 107, 113

epsom salts .... *See* magnesium sulphate

Equipment

    beaker 143, 144, 154, 157, 159, 160, 163, 164, 233

    blender 142

    borosilicate 154

    brushless drill ... 147, 154, 161

    bucket 120, 146, 149, 161, 162, 163, 165, 166, 169

Buckner flask............. 231
cell-culture dish 144
conical flask....... 144, 235
cooking pot140, 146, 154, 155, 167, 210, 235
cylindrical beaker 144
demijohn................... 193
drum . 125, 150, 179, 180, 185, 186, 187, 192, 213
Erlenmeyer flask 144
eye dropper.............. 142
flexi tub..................... 168
flexi tubs 150, 168
fume cabinet...... 110, 151
funnel 149, 166, 167, 231
Funnel 231
gas chromatograph ... 233
graduated cylinder ... 144, 163
hardstand 190
home brew buckets.. 146, 154, 164, 169
hotplate 145, 154, 155, 156, 160, 167, 235
hydrometer143, 154, 163, 225, 232
in-line filter 149, 166, 167, 191, 225
kilner.......... 147, 154, 182
lab spoons................ 147
laboratory beaker 185, 233
ladle... 149, 162, 164, 165
litmus 144, 164, 166, 206, 231, 251, 253

maslin pan 146, 154, 155, 167, 210, 235, 236
measuring cylinder.... 144
micro spoons..... 147, 154
mixing paddle ... 148, 154, 161
Pasteur pipette .. 142, 157
Petri dish 144
Petri plate 144
pH meter.... 144, 166, 206
rods 144
Sandy Brae test flask 236
scales141, 154, 221, 222, 223, 224, 235
scoops...................... 147
siphon pump............. 141
spatulas.................... 147
spill kit...................... 154
stopwatch .. 148, 155, 163
tank..... 23, 24, 33, 85, 86, 87, 88, 170, 179, 180, 181, 182, 183, 184, 187, 190, 191, 192, 193, 194, 197, 201, 203, 217, 219, 221, 225, 226, 233, 246, 254, 256
test sieves . 145, 169, 170
thermometer..... 143, 154, 155, 167, 168, 214, 230, 235
thermostat 192
transfer pump 141
vacuum pump............ 231
Zahn cup viscosimeter
.... 148

erucic acid 54, 246

ESIS 96, 108, 114, 117

ester...50, 51, 91, 174, 248, 250, 252, 259

esterification 91, 248

ethane 45, 47, 49

ethanoic acid ..... See acetic acid

ethanol. 49, 50, 62, 91, 106, 248, 251, 257

ethene 47

ethyl acetate .... See ethyl ethanoate

ethyl ester 248

ethyl ethanoate.............. 51

ethylene.......... 47, 250, 251

ethylene oxide 106

ethyne........................... 47

Euro 76

Europe....41, 55, 59, 60, 82

exempt producer . 174, 175, 177

exothermic.................... 248

expression 66

FAME's..52, 53, 63, 65, 67, 74, 75, 92, 93, 152, 247, 248, 249, 253, 257

fatty acid methyl esters . 52, 92, 93, 152, 248, See FAME's

feedstock 34, 72, 73, 249

ferric-oxide.................... 66

ferric-oxide polymer......See ferric oxide

FFA 93, 153, 156, 157, 158, 162, 170, 179, 199, 206, 222, 224, 231, 241, 242, 243, 245, 248, 249, 255, 256

FIA Formula E 106

Fiat

   1400-A........................ 17

   Croma TD-i.d............... 19

Finland............................ 64

flashpoint 83, 249, 259

fluorine.......................... 249

fluorocarbon.................. 249

fluoropolymer........ 255, 258

fluroelastomer 89, 149, 187, 191, 226

Ford 55

   Model T........................ 12

   soybean car................. 55

formaldehyde............ 72, 95

formic acid 50, 95, 248

fossil fuels.... 13, 34, 58, 67, 250

four stroke........... 12, 20, 21

fractional distillation .. 36, 37

France 11, 13

free radicals . 199, 200, 224, 245, 249

fuel filter heater............. 217

fuel tank heater............. 217

gel point... 52, 53, 212, 213, 214, 215, 217, 230, 249, 254, 255

Germany 11, 110

GHS 97, 99, 102, 104, 108, 112, 114, 117

glow plugs 20, 218

glycerin .... *See* glycerol

glycerine .... *See* glycerol

glycerol 75, 91, 92, 93, 105, 152, 156, 161, 162, 163, 164, 168, 172, 193, 202, 205, 206, 207, 208, 209, 210, 211, 219, 220, 221, 222, 223, 227, 228, 233, 239, 240, 245, 247, 249, 250, 253, 257, 259

Glycine max .. *See* soybean

GM.................................. 55

Goldman Sachs............. 61

greenhouse gas . 69, 70, 71

haemoglobin.................. 70

haloalkanes 48

hard water 246, 252

HDPE 96, 107, 186, 187, 190, 250

heat exchanger 86, 88

   coaxial 86

   flat plate...................... 86

Helianthus annus .... *See* sunflower

helix controlled direct injection 25

hemp 55, 72, 74, 75, 215

heneicosane.................. 45

Henry Ford 55

hentriacontane 46

heptacosane.................. 45

heptadecane 45

heptane 45

heroin 114

hexacosane................... 45

hexadecane................... 45

hexane..................... 45, 66

High Melting Point Esters 230, 231

HM Revenue & Customs 175

Honda...................... 19, 55

   Accord 19

   i-CTDI 19

HSE 177, 178

humectant 209

Hungarian.................... 110

hydrocarbons 28, 35, 44, 46, 47, 48, 64, 68, 71, 79, 81, 250, 254

hydrochloric acid . 113, 114, 115, 204, 206, 255

hydrogen 13, 14, 35, 43, 44, 47, 48, 64, 71, 99, 101, 103, 236, 245, 250, 255

hydrogen acetate .... *See* acetic acid

hydrogen chloride......... 113

hydrogen sulphate........*See* sulphuric acid

hydrogenated .. 52, 64, 213, 250, 257

hydrogenation. 52, 250, 253

hydrolyse 199, 248

hydrolysis 199

hydronium chloride.......*See* hydrochloric acid

hydroperoxide....... 199, 249

hydrophilic 250

hydroxyl 48, 50, 248

hygroscopic .. 184, 198, 250

IDI 18, 239

immiscible.................... 250

India............................. 12

indicator solution 251

indirect injection. 18, 23, 24, 25

individual fuel system 23

Indonesia....................... 58

industrial revolution 70

infrared 69, 143

injection nozzle............. 247

injectors23, 24, 25, 29, 222, 224, 226, 245

insulation 191

intercooler..... 22, 27, 77, 78

internal combustion engine 11, 12, 14, 15, 19, 28

iodine 259

IPA....*See* isopropyl alcohol

iron.................. 88, 103, 191

isocyanates................... 106

isopropanol... *See* isopropyl alcohol

isopropyl alcohol.. 106, 107, 108, 109, 156, 157, 234, 251, 255

isosane 45

Isuzu

   Bellel............................ 17

Iveco 55

Japan....................... 12, 55

jatropha................... 61, 72

Jatropha curcas ....*See* jatropha

Jean Joseph Etienne Lenoir.............................. 11

kerosene....... 34, 35, 36, 37

kinematic viscosity........ 251

KOH standard solution 153, 157, 158, 170, 234

lambda........................... 33

Lancia 28

   Delta S4....................... 28

LDPE 251

Le Mans......................... 19

lead........................ 99, 101

Legislation

   Alternative Fuels Infrastructure Directive

2018' (EU Directive 2014/94/EU) 40

Euro 5b 77

Euro 6 77

Excise Notice 179e: biofuels and other fuel substitutes 175

Integrated Pollution Prevention & Control Licence 177

marihuana tax act 55

Measuring Instruments Directive 2014/32/EU 178

Measuring Instruments Regulations 2016 ... 178

Personal Protective Equipment at Work Regulations 1992 ... 136

The Alternative Fuel Labelling and Greenhouse Gas Emissions (Miscellaneous Amendments) Regulations (2019)... 40

The Control of Pollution (Oil Storage) (England) Regulations 2001 ... 174

Transport Biofuels Directive 54

United Nations Convention Against Illicit Traffic in Narcotic Drugs & Psychotropic Substances 114

Waste Carriers Licence 177

Weights & Measures . 178

linolenic acid 251, 259

lipids 91, 255, 257

LLDPE 251

LNT 79

Lord Howard de Clifford . 17

Lotus 55

LPG 37, 39

LSD 42, 81

lubricant 22, 41, 56, 251

Ludwig Elsbett 90

lye207, 251, See potassium hydroxide

Macedonia 41

Madagascar 58

magnesium 103, 252

magnesium stearate 252

magnesium sulphate 204

maize .... See corn

Malaysia 58

manganese 103

meniscus 143, 163

Mercedes 17, 55

methamphetamine 114

methane 45, 48

methanecarboxylic acid See acetic acid

methanol48, 50, 91, 92, 93, 95, 96, 97, 98, 121, 126, 127, 128, 137, 138, 141, 145, 147, 155, 159, 160,

161, 162, 164, 165, 166, 168, 169, 179, 180, 182, 184, 186, 187, 193, 194, 202, 203, 205, 207, 209, 210, 219, 220, 222, 223, 229, 232, 233, 238, 239, 240, 241, 242, 243, 246, 248, 252, 254, 256, 257, 259

methoxide.... 159, 160, 161, 171, 190, 191, 192, 193, 194, 239, 240, 252, 254

methyl.....50, See methanol

methyl acetate.. See methyl ethanoate

methyl alcohol ....See methanol

methyl cyanoacrylate ... 253

methyl ester.......... 252, 259

methyl ethanoate........... 51

methyl formate.. See methyl methanoate

methyl methanoate. 50, 248

Mexico 60, 62

Middle East..................... 65

mild steel 88, 96, 107

miscibility 252

Mitsubishi 55

molecular sieve 252

Mongolia...................... 216

monoglyceride 152, 181, 184, 222, 229, 247, 253, 257, 259

monounsaturated fat 52, 56, 60, 253, 257

Monte Carlo Rally 17

muriatic acid ....See hydrochloric acid

naphthenes................... 35

national debt............ 83, 84

Neste Oil......................... 64

Netherlands 64

New Rational Combustion Engine 12, 15

New Zealand 65

nickel 103

nicotine 208

night heater.................. 218

Nikolaus August Otto 11, 12

nitric oxide 32, 71

nitrile 253

nitrogen.. 32, 33, 35, 48, 66, 68, 69, 71, 78, 79, 138

nitrogen dioxide .. 32, 33, 71

nitrogen oxide.... 32, 33, 63, 68, 71, 74, 75, 76, 78, 79, 80, 81, 82

nitrous oxide 71

nonacosane.................. 46

nonadecane.................. 45

nonane.......................... 45

North America 55, 59

Northern Canada.... 63, 216

..... 216

.... 19

...cosane 45

octadecane.. 45

octane 38, 45, 247

octane number 38

oil crisis 12

oil of vitriol .... *See* sulphuric acid

oil palm 58, 66, 72

OPEC 34

oxidation 31, 120, 233, 245, 248, 259

oxidative polymerisation. 88

oxidisation 199

oxidise 28, 199, 249

oxidised 70, 199, 233

oxygen... 19, 28, 32, 33, 35, 43, 44, 48, 49, 68, 69, 70, 71, 72, 127, 138, 199, 200, 207, 226, 245

ozone 71

PAH 82

Pakistan 61

palladium 28, 32

PAPR 95, 107, 122

paraffin 35

Paris 53

...ually hydrogenated. 213, 253, 257

particulate matter 28, 31, 68, 72, 76, 77, 81, 145

peak oil 83

peanut 15, 53, 72

pentacosane 45

pentadecane 45

pentane 45

pentatriacontane 46

peroxide *199*, 249

petrochemical diesel *40*, 41, 56, 63, 64, 65, 68, *73*, *75*, 76, 79, 80, 81, 82, *83*, *84*, 86, 88, 90, 150, 155, *163*, 164, 174, 212, 213, *217*, 225, 226, 245, 257

Peugeot 18, *89*

  205 .... *18*

  305 .... *18*

  405 .... *18*

pH 111, 114, 116, 144, 145, 166, 182, 183, 204, 206, 231, 245, 251, 253, 254

phenol red 253

phenolphthalein... 110, 111, 154, 157, 234, 251, 254

  Adolf von Bayer 110

  Ex-Lax 110

  Max Kiss 110

Philippines 57, 61

pHLip 233

phospholipids 64

phosphorus....... 48, 66, 259

photobioreactors............ 6

photosynthesis 65

piezoelectric injector..... . 24

piston 20, 25, 78, 247

platinum............... 28, 32

plunger 23, 24

PM 76, 81

polyethylene 142, 183, 250, 251, 254

polymer......... 250, 254, 255

polypropene................ See polypropylene

polypropylene 254

polyunsaturated fat.. 55, 56, 60, 74, 75, 254, 257, 259

polyvinyl chloride.......... 255

polyvinylidene fluoride.. 255

Porche 55

potash lye ... See potassium hydroxide

potassia ...... See potassium hydroxide

potassium 254

potassium hydrate........ See potassium hydroxide

potassium hydroxide 93, 101, 102, 119, 140, 153, 154, 158, 159, 170, 203, 204, 205, 206, 207, 209, 210, 237, 238, 239, 241,

potassiu...

pour point............

PPE .10, 94, 121, 122, 12.. 129, 155, 202, 204, 206

BA.............. 136, 137, 138

breathing apparatus . 133, 134, 135, 136, 137, 138

coveralls ... 122, 132, 133, 154

eye wash .. 119, 125, 126, 154

face protection... 136, 154

face shield ... 95, 107, 136

fire blanket......... 128, 154

fire extinguisher 126, 127, 128, 154

first aid kit 126, 154

gauntlets... 122, 129, 130, 131, 154

gloves .95, 107, 110, 120, 122, 129, 130, 131, 132, 154, 253

goggles 95, 107, 136, 154

HAZMAX.................... 135

hose pipe .. 119, 127, 128, 140, 154, 161, 180

mobile phone.... 138, 140, 154

NCB 122, 133

respiratory protection equipment.............. 136

RPE 136, 137

safety spectacles....... 136

shower head..... 127, 128, 140

spill kit 129

thermal imaging. 139, 140

vinegar 50, 116, 117, 140, 154

visor.......................... 136

welding blanket 128

Wellington boots 122, 123, 134, 135, 154

propan-2-one.. *See* acetone

propane 39, 45, 49, 245

propanetriol .... *See* glycerol

propanol 49, 50

propanone ...... *See* acetone

propionic acid 50, 51

PTFE 187

PVC.............. 191, 247, 255

PVDF........................... 255

Rancimat test 200, 255

rapeseed 54, 60, 72, 74, 75, 215, 246, 255

Rational Combustion Engine ... *See* New Rational Combustion Engine

RCD 121

regeneration 79

renewable diesel 63, 64

rhodium 32

RON 38

rotary fuel system........... 23

Rudolph Diesel.. 12, 15, 16, 53, 91

safflower 56, 72, 73, 74, 75, 215, 254

Samoa 57

saponification 255

saturated fat 52, 53, 54, 55, 56, 57, 60, 61, 63, 89, 213, 215, 255

SCR........ 32, 33, 78, 79, 80

sec-propyl alcohol .... *See* isopropyl alcohol

Serbia 41

silicate 72

silicon carbide 29, 31

Singapore 64

smog............................. 71

smoke point.................. 255

soda lye .... *See* sodium hydroxide

sodium carbonate......... 246

sodium hydroxide 93, 98, 99, 100, 101, 119, 159, 205, 206, 207, 209, 210, 220, 237, 238, 239, 240, 241, 242, 243, 244, 245, 246, 247, 250, 251, 252, 253, 254, 256

sodium methoxide 256

sodium stearate........... 256

solenoid 25, 88

South America.... 53, 62, 82

South Korea 12

South Pacific 57

Southern Canada 62

soybean . 55, 58, 59, 60, 61, 63, 72, 254

spark plugs 20, 22, 38

specific gravity 163, 225, 232

spirits of salt .... *See* hydrochloric acid

spontaneous combustion 120

Sri Lanka 58

stainless steel .. 88, 96, 107, 145, 146, 147, 149, 154, 155, 167, 186, 190, 210, 235

Standard

   Vanguard 17

Standards

   ANSI Z 87.1 136

   ASTM D6751 200

   BS718 143

   EN1146 138

   EN12662 259

   EN12941 137

   EN12942 137

   EN13034 .... 132, 135

   EN137 137

   EN13832 134

   EN14103 259

   EN14104 259

   EN14105 259

   EN14106 259

   EN14107 259

   EN14108 259

   EN14109 259

   EN14110 259

   EN14111 259

   EN14112 259

   EN14214 ... 174, 200, 235, 259

   EN145 138

   EN14605 .... 132, 135

   EN166 136

   EN1835 137

   EN1869 128

   EN270 137

   EN374 131

   EN388 .... 129, 130

   EN402 138

   EN533 133

   EN943 .... 132, 134, 135

   ISO11611 133

   ISO11612 133

   ISO12185 259

   ISO12937 259

   ISO13982 135

   ISO139872 132

   ISO13997 .... 129, 130

   ISO14116 133

   ISO20344 135

   ISO2160 259

   ISO310 259

   ISO3675 259

   ISO3679 259

   ISO3987 259

   ISO5165 259

stearic acid ... 252, 254, 256

steel 88, 119, 190

sulphate 72

sulphur . 35, 48, 64, 82, 259

sulphur dioxide 82, 104

sulphur oxide 79

sulphuric acid 103, 104, 105, 119, 123, 243

sunflower 60, 72, 254

supercharger 27, 28, 78, 80

    Centrifugal 27

    Roots type 27

    Twin screw 27

SVO. 18, 52, 60, 72, 75, 76, 79, 80, 82, 85, 86, 87, 88, 89, 90, 93, 152, 153, 154, 156, 157, 158, 159, 160, 162, 163, 164, 171, 172, 175, 213, 214, 229, 234, 249, 252, 255, 256, 257

Switzerland 41

Talbot

    Horizon 18

TDC 26

TEFC pump 193

Tesla 217

test sieves 224, 225

test solution 152, 153

tetrabromophenolsulfonphthalein ...... See bromophenol blue

tetracosane 45

tetradecane 45

tetratriacontane 46

thermoplastic244, 247, 250, 251, 254, 255

tin 99, 101

titrates .... See titration

titration... 93, 153, 156, 157, 158, 159, 162, 170, 222, 223, 224, 234, 237, 238, 239, 240, 241, 243, 244, 255, 256

titrations .... See titration

Trading Standards 178

trans fat .... 253, See hydrogenated oil

trans fatty acid 257

transesterification 61, 91, 92, 93, 159, 162, 223, 225, 240, 243, 244, 246, 248, 252, 256, 257

    base-catalysed 91

    E Duffy 91

    G Chavanne 91

    J Patrick 91

triacontane 46

tricosane 45

tridecane 45

triglyceride . 91, 92, 93, 257, 259

tritriacontane 46

TSI 28

turbocharger 26, 27, 28, 78, 80

Turkey 41, 56

Tutankhamen 56

twincharger.................... 28

UK.... 13, 14, 38, 40, 41, 62, 83, 84, 96, 99, 101, 104, 108, 110, 114, 117, 123, 173, 176, 177

ULSD 42

UN 186

undecane....................... 45

unsaturated fat .. 52, 61, 74, 75, 89, 199, 249, 254, 257

urea 32

USA . 13, 41, 42, 53, 54, 55, 56, 59, 62, 63, 64, 65, 97, 99, 102, 104, 108, 114, 117, 200

vacuum pump....... 203, 231

valency 44

Vanuatu 57

viscosity . 37, 38, 64, 75, 86, 88, 148, 163, 164, 171, 257, 259

Viton 89, 142, 149, 187, 191, 226, 258

VOC............................... 81

Volkswagen 28, 55

emission scandal......... 77

Jetta TDI 79

Tiguan......................... 79

Volvo............................. 55

S60 28

S90 28

XC60........................... 28

XC90........................... 28

West Africa 58

West Indies.................... 58

wood alcohol See methanol

WVO 39, 75, 82, 84, 90, 93, 145, 152, 155, 156, 169, 170, 171, 172, 175, 177, 206, 213, 214, 234, 235, 236, 238, 239, 240, 243, 245, 249, 251, 252, 253, 254, 255, 256, 257, 258

WWI.............................. 12

WWII........... 53, 55, 59, 209

XUD 18, 89, 90

Zea mays.............. See corn

zeolite 79

zinc 99, 101, 103

About The Author

Mr Xavier is in his 40's & currently living in the South of England. He is desperately trying to make a living, live a good life, learn to play the ukulele, grow a rather splendid handlebar moustache & decide what to have for dinner tonight. With his other hand he's trying frantically to learn another language whilst also planning the next chapter of his life.

This & all his other books are available online from reputable book & ebook retailers.

Made in United States
Troutdale, OR
09/21/2024

23011656R00155